The Third Nuclear Age

The Third Nuclear Age

Complexity, Contestation, and the Evolution of Global Order

ANDREW FUTTER

OXFORD
UNIVERSITY PRESS

OXFORD
UNIVERSITY PRESS

Great Clarendon Street, Oxford, OX2 6DP,
United Kingdom

Oxford University Press is a department of the University of Oxford.
It furthers the University's objective of excellence in research, scholarship,
and education by publishing worldwide. Oxford is a registered trade mark of
Oxford University Press in the UK and in certain other countries.

Published in the United States of America by Oxford University Press
546 Fifth Avenue, New York, NY 10036, United States of America

British Library Cataloguing in Publication Data
Data available

Library of Congress Control Number: 2026937764

ISBN 9780198944096

DOI: 10.1093/9780198944126.001.0001

Printed and bound by
CPI Group (UK) Ltd., Croydon, CR0 4YY

MIX
Paper | Supporting
responsible forestry
FSC® C013604

The manufacturer's authorized representative in the EU for product safety is
Oxford University Press España S.A. of Parque Empresarial San Fernando de Henares,
Avenida de Castilla, 2 – 28830 Madrid (www.oup.es/en or product.safety@oup.com).
OUP España S.A. also acts as importer into Spain of products made by the manufacturer.

For Rehana

Acknowledgements

I have spent the best part of 20 years thinking about global nuclear politics, but this book simply would not have been possible without four things.

The first was the generous grant from the European Research Council (ERC) that I was awarded in 2020, and which allowed me the time, resources, and most importantly the ability to build a team to think about how to understand our changing nuclear world. This in turn was only made possible by the constant support of various different people at the University of Leicester who helped with the initial proposal and walked me through the peculiarities of ERC procedures and reporting, and in particular Steff Collins and Krupesh Mistry who unfailingly answered all my silly questions and tried to keep up with my various 'hair-brained' plans.

The second, is the brilliant people that I have been surrounded by while undertaking this project and organising my thoughts on what a Third Nuclear Age is, might mean, and how it should be conceptualised. The wonderful team I was able to build at Leicester was at the heart of this: Ludovica Castelli, Cameron Hunter, Olamide Samuel, Shivani Singh, and Francesca Silvestri have all had an enormous impact on the way I think about the nuclear world. So too have my good friends Bleddyn Bowen, Benoit Pelopidas, Nick Wheeler, and Ben Zala, all of whom I always secretly suspected knew more about the topic than I did, and the brilliant group of scholars that were part of the 'Nukes-Strat-Space' group at Leicester.

Third is the fantastic network of very smart, generous, and engaging people that I have either known for a while or have met as part of this research, from academic, professional, and policy worlds. I have been so lucky to meet so many people, in so many parts of the world that want to engage and find a way of reducing the dangers that we have lived with since 1945, and give time for people like me that want to have a stab at trying. There are many people that I probably can't name here, and possibly too many without risking an Oscar's speech, but hopefully you know who you are. A special thank you is due to those hardy souls who travelled to Leicester on the hottest day of 2025 and took part in the book workshop.

Last, and perhaps most importantly, are those people who had to live with me as I jetted off to a conference, fretted about some incomprehensible detail, scattered my books and papers all over the house, or was generally lost in some daydream, as I worked on the project and this book. For your forbearance, support, and love, thank you Rehana and Zane.

Contents

Introduction

Welcome to the Third Nuclear Age

Midway through the third decade of the 21st century we stand at an inflection point in our nuclear history.[1] After what seems like a generational hiatus where for many people the world appeared to become more globalised, liberal, and peaceful, nuclear weapons are firmly back as a central currency of international politics and as a reality of everyday life. Fears of nuclear use and nuclear war that many—especially in North America, Europe, and 'the West'—hoped had receded into the past have returned, and nuclear weapons have reappeared as the ultimate symbols of a hypermasculine expression of state power. Above all, there is a growing concern that we are sleepwalking into a new nuclear era for which humanity is ill prepared as a species. As United Nations (UN) Secretary General Antonio Guterres warned in September 2024:

> Not since the worst days of the Cold War has the specter of nuclear weapons cast such a dark shadow. Nuclear saber-rattling has reached a fever pitch. We have even heard threats to use a nuclear weapon. There are fears of a new arms race. Meanwhile, the norms painstakingly established over decades against the use, spread and testing of nuclear weapons are being eroded.[2]

A decade and a half after United States (US) President Barack Obama stood on a podium in the Czech capital Prague and championed his 'commitment to seek the peace and security of a world without nuclear

[1] This section heading is a title also used by Karl Heinz Kamp in 'Welcome to the Third Nuclear Age', *National Interest*, 2 May 2016, https://nationalinterest.org/feature/welcome-the-third-nuclear-age-16020.

[2] 'Secretary-general's remarks at high-level meeting commemorating and promoting the international day for the total elimination of nuclear weapons', 26 September 2024, https://www.un.org/sg/en/content/sg/statement/2024-09-26/secretary-generals-remarks-high-level-meeting-commemorating-and-promoting-the-international-day-for-the-total-elimination-of-nuclear-weapons-bilingual-delivered-scroll.

The Third Nuclear Age. Andrew Futter, Oxford University Press. © Andrew Futter (2026).
DOI: 10.1093/9780198944126.003.0001

weapons',[3] it appears that the human story has moved in a different and far more worrying direction. In fact, we are perhaps as far away from the goal of nuclear disarmament today than at any point since Ronald Reagan and Mikhail Gorbachev sat down to talk inside Höfði House in Reykjavik in October 1986, and possibly since the 'gadget' exploded into the New Mexico desert sky early in the morning of 16th July 1945.

The reality is that we have arrived in a distinctly different material and ideational chapter in our shared nuclear story, characterised by political, technological, normative, and intellectual change, complexity, and contestation. At its heart is the transition from a nuclear world of US and Western dominance, coherence, and power, a sustained period of nuclear weapons reductions, a focus on non-proliferation and non-state actor threats, and a general public apathy towards anything nuclear, to a world of strategic competition, a gradual move towards *nuclear* multipolarity and competition, arms build-ups, renewed interest in civilian nuclear technologies, growing vocal frustration at the efficacy and fairness of nuclear governance frameworks, and a more sustained challenge to the logic of nuclear eternalness.[4] While much has been written about the considerable geopolitical and cultural shifts that accompanied the end of the Cold War and how they impacted the global nuclear order, the change we are living through today is different and perhaps more diverse. This has implications for what we might consider as nuclear orthodoxy—the set of values, beliefs, and concepts that have become normalised over time and shape our understanding of the nuclear condition. Taken together, we can think of this as a new and distinct Third Nuclear Age.

Questions abound in this Third Nuclear Age: what has (and hasn't) changed from the past and how might this impact the way that we think about nuclear risks? Who decides this narrative, does anyone seek to gain from it? How might 'exotic' 'disruptive' technologies such as 'hypersonic' missiles, autonomous platforms, space weapons, and artificial intelligence (AI) impact deterrence, stability, and the likelihood of inadvertent escalation? Will great power nuclear competition between the US, Russia, and

[3] 'Remarks by President Barack Obama in Prague as delivered', The White House Office of the Press Secretary, 5 April 2009, https://obamawhitehouse.archives.gov/the-press-office/remarks-president-barack-obama-prague-delivered. As is noted later in the book, the speech also contained a number of caveats.

[4] The idea of nuclear eternity comes from the French intellectual and scholar, Benoit Pelopidas. See Benoit Pelopidas, 'The birth of nuclear eternity', in Sandra Kemp & Jenny Andersson (eds.), *Futures* (Oxford: Oxford University Press, 2021).

China result in an increased chance of interstate nuclear war? What does this mean for alliance systems and the role and influence of a variety of influential 'middle powers'? Can the world move back to arms control and restraint, what does this look like, and who benefits from this—or is our nuclear future more likely to be characterised by proliferation, shocks, and antagonism? What is the short-term future of the push for nuclear disarmament; is it on life support, can we survive without it, and what is the best mechanism to achieve it? Can peaceful nuclear technologies spread both to multinational companies to fuel cutting-edge science and across the non-Western world to combat energy, net zero, and developmental demands, without weaponisation or increases in other security threats? And how do we make sure that policymaking elites and the public have the necessary understanding and tools to engage with the nuclear-climate-development nexus that sits at the heart of what has become termed the 'Anthropocene'?

This book is both a natural successor to the rich literatures on the intellectual constructs of a First and Second Nuclear Age and at the same time a direct challenge to the (understandably) focussed, often Western ethnocentric, and in some ways instrumental and constrained view of much of that literature.[5] This is because it deliberately seeks to move beyond the inherent narrative focus on the largely English-speaking, Euro-Atlantic world, the nuclear-armed states and those that might seek to acquire nuclear weapons, to engage with all aspects of nuclear order and how we think about the broader impact of nuclear technologies and nuclear imaginaries. This includes seeking to engage with the views and aspirations of different nuclear stakeholders, and across different levels of participation in nuclear politics (not just 'states', but policy elites, influential non-governmental organisations (NGOs), academic debate, and public perceptions) within the international system, and interacting with non-Anglophone thinking and scholarship. It also involves treating 'nuclear' as a wider concept that includes not just weapons but other possibly 'nutopian' applications as well.

A key underlying theme of this book is that different entities have different views and understandings of the notion of nuclear order, how it should be periodised, what it should look like, and how it should work: it is inherently contested. But for the majority of the nuclear epoch[6] —or at least the period since 1945—nuclear order, and the way that we think about nuclear

[5] This literature is addressed in detail in Chapter 1.
[6] As Nicholas Wheeler referred to it in a recent conversation.

weapons, has reflected the overwhelming dominance of the US on international politics. The argument here is not that this three-quarter century nuclear *pax Americana* has come to an abrupt stop, but that it is becoming slowly more nuanced as power balances and intellectual paradigms begin to shift globally. Naturally, this is playing out within a classic structure vs agency, context vs conduct environment, within which certain actors have more ability, influence, and power than others to drive, embrace, or resist this change. Ultimately however, the book argues that those with the most power will continue to shape nuclear politics in this new era.[7] Whether these so-called great powers will act 'responsibly' remains to be seen.

For some policy, diplomatic, and military elites, the Third Nuclear Age is about doubling down on nuclear weapons and deterrence (albeit in different ways), for others it is about seeking to promote restraint and stability, and protecting established global nuclear frameworks, norms, and obligations, while for others still it is about pushing for a fairer and more just 'grand bargain' on the road to disarmament and a pathway to sustainable economic development.[8] Thus, at the heart of the Third Nuclear Age is a battle between those committed to maintaining a slightly modified status quo and those pushing for different types of transformation and revisionism. We can think of these as different manifestations of 'nuclear friction', whereby technological, geopolitical, normative, and intellectual drivers of change will clash, possibly in unpredictable ways. The result is what the prominent Indian scholar Manpreet Sethi describes as a more permissive international nuclear environment, both materially and ideationally.[9]

The main objective of *The Third Nuclear Age* is therefore to map out and conceptualise this ongoing change in our global nuclear world, and seek to understand the major fault lines, challenges, and points of disagreement, so that we can prepare ourselves properly for the nuclear future. A key part of this is re-engaging with our shared—but often differentiated—global experience of nuclear history. The book is therefore deliberately written to start a debate and challenge entrenched and existing viewpoints

[7] Perhaps the best way to think about power in nuclear/international politics is captured in Michael Barnett & Raymond Duvall, 'Power in international politics', *International Organization* 59:1 (2005), pp. 39–75. As the authors put it, 'Power is the production, in and through social relations, of effects that shape the capacities of actors to determine their circumstances'.

[8] These ideas are also discussed in Andrew Futter, Ludovica Castelli, Cameron Hunter, Olamide Samuel, & Francesca Silvestri, *The global Third Nuclear Age: Clashing visions for a new era in international politics* (London: Routledge, 2025).

[9] Discussion with the author.

and shibboleths and add a different perspective to the embryonic and rich literature analysing the 'New' or 'Third' Nuclear Age.

Unlike the collapse of the Soviet Union in 1991 and the end of the Cold War that marked a shift from the First to the Second Nuclear Age, there is no obvious date or event that marks the shift into a Third Nuclear Age. At least part of the reason is that this historical periodisation only really makes sense if you are sitting in North America or Europe—China joining the nuclear club in 1964, the 'Vela incident' of 1979,[10] the nuclearisation of south Asia in 1998, or the North Korean nuclear test of 2006, as well as other moments are arguably major inflection points for other states that constitute the global nuclear order. For most people across the globe, the nuclear sword of Damocles that arrived unwanted in the 1950s with the advent of thermonuclear weapons and long-range delivery platforms, and subjected everyone across the world regardless of gender, faith, or nationality to the stark reality of human-driven extinction, hasn't changed much since.

But the reasons for the change that has occurred over a period of about a decade roughly between the early 2010s and the early 2020s are significant, this is because what animates the focus of the most dominant players within the global nuclear ecosystem continues to frame the way that global nuclear politics is conceptualised. Under this—admittedly bounded—view the nuclear centre of gravity slowly shifted away from the threats posed by rogue states, loose fissile material, and terrorists towards the challenges of great power competition and the possibility of interstate war in a more contested international system. That said, in stark contrast to earlier framings, this shift also appeared much more globalised, and far less centred on how the *nuclear* world looked from Washington, DC.

This shift has not necessarily been linear or coherent, as nuclear politics ebbed and flowed in something of a punctuated equilibrium, and as the Second Nuclear Age slowly merged into the Third. Instead, we saw change being driven at different speeds and with different implications across three broad categories: technological innovation and diffusion in weapons, support systems, and context, (and increased spending to support this); the distribution of political power within the international system moving gradually away from US dominance towards increased nuclear multipolarity, and at the same time an erosion of the perceived legitimacy (and coherence) of the Western liberal international project; and a growing divide

[10] The suspected joint test by Israel and South Africa of a nuclear device in the southern Indian Ocean, picked up by a US 'Vela' satellite.

in approaches to nuclear security, disarmament, development, and visions for the nuclear future within this context. The agreement of the Treaty on the Prohibition of Nuclear Weapons (TPNW) at the UN in 2017, the end of the Intermediate-Range Nuclear Forces (INF) Treaty in 2019, the modernisation and expansion of nuclear arsenals by all nuclear-armed states, the return of overt nuclear threat making and 'first-use' doctrines by the leaders of certain states, the inability to reach consensus at either of the two recent Treaty on the Non-Proliferation of Nuclear Weapons (NPT) review conferences (at the time of writing the 2026 RevCon hadn't taken place), the panic about US disengagement and renewed concerns about 'nuclear dominos' in Europe, northeast Asia, and the Middle East, a growing demand for nuclear technology in the developing world, and an amplification of previously marginalised viewpoints during this period, all point to notable changes from the generation that preceded it.

Our nuclear world today looks far more dangerous than it did in in the mid-1990s when South Africa unilaterally disarmed and the international community came together and agreed to indefinitely extend the NPT; in the mid-2000s as thousands of anti-nuclear-war protestors marched in New York City, the Treaty of Bangkok established a nuclear-weapons-free zone in Southeast Asia, and the '4 horseman of the nuclear apocalypse'[11] called for nuclear disarmament; or even in the early 2010s as the US and Russia 'reset' their nuclear relationship, a deal was struck with Iran to limit its nuclear programme, and major stakeholders came together for the Nuclear Security Summits to secure fissile material and safeguard nuclear facilities.

Today there are fears of conflict between major powers escalating to nuclear use—particularly, but not limited to, the war in Ukraine and eastern Europe, the status of Taiwan, and the strategic dominance of the south China Sea, and simmering conflicts on the Korean peninsula, in south Asia, and across the Middle East; the return of 'tactical' battlefield nuclear forces and nuclear coercion; fears of nuclear weapons being used in space; a suite of exotic military technologies that may fundamentally challenge established ideas of strategic stability and deterrence; a revitalisation of the existential link between nuclear technologies, societal development, scientific progress, and climate change; and a major and growing split between two visions of our future—one that requires a system of managed nuclear deterrence and one that demands immediate nuclear disarmament.

[11] Former US Secretaries of State Henry Kissinger and George Shultz, Defense Secretary William Perry, and former Senator Sam Nunn.

At the same time, this book doesn't seek to claim that everything is new. Fears that dominated the nuclear debate in the 1990s and 2000s have not disappeared; the risks of non-state actors gaining access to nuclear material or nuclear facilities remains. The same is true for the worries that shaped the 1960s, as the fear of nuclear proliferation beyond the nine states that currently have nuclear weapons intensifies. These anxieties have merely been supplanted by additional nuclear concerns—they are not historical problems that have been solved. In this way, the Third Nuclear Age can be thought of as a combination of both First and Second Nuclear Age pressures. Or as *New York Times* journalist David E. Sanger put it—the new geopolitical era 'is part 1914, part 1941, and part 2022. All at once'.[12] This is why the Third Nuclear Age conceptualisation used here allows for change to occur at different times, in different ways, and across different dynamics.

Perhaps the greatest fear in the Third Nuclear Age is that we humans are ill prepared to engage with this set of complex challenges. A long post-Cold War peace dividend, especially in the Western world has allowed a certain complacency, apathy, and ignorance to grow, whereby nuclear risks have been displaced by other types of challenge to society. In the 1980s almost everyone in Europe and North America worried about the threat of nuclear war, and vast numbers of people engaged in anti-nuclear protests and demonstrations, and in some cases were ordered to carry out nuclear alert drills. In 1984, BBC2 in the United Kingdom (UK) aired the film *Threads* depicting the horrifying aftermath of a fictional nuclear attack on the UK city of Sheffield. A year earlier the ABC network in the US had depicted similar gruesome outcomes of a nuclear attack on small town (called Lawrence) in the state of Kansas in *The Day After*, and Hollywood introduced cinema goers to the Pandora's box that could result from the interaction between computers, AI, and nuclear weapons in the movie *WarGames*, and a year later in *The Terminator*. These films, as well as books such as General Sir John Hackett's *The Third World War*[13] depicting in great detail a nuclear conflict between East and West, and the powerful animated film *When the Wind Blows* captured the fears of populations across particularly the Euro-Atlantic world and reflected the very real anxiety of imminent and devastating nuclear destruction. Indeed, in 1986 Konstantin Lopushansky

[12] David E. Sanger, *New cold wars: China's rise, Russia's invasion, and America's struggle to defend the West* (Melbourne: Scribe, 2024), p.15.

[13] General Sir John Hackett, *The third world war: A future history* (London: Sidgwick & Jackson, 1978). Hackett also published a sequel several years later seeking to unpack what had happened: *The third world war: The untold story* (London: Sidgwick & Jackson, 1982).

directed *Dead Man's Letters* (*Písma myórtvogo chelovéka*) the first Russian film to explore the aftermath of nuclear war in the Soviet Union.

But beginning with the end of the Cold War, nuclear anxieties began to fall away, and American scholar Francis Fukuyama's notion of an end of history began to take hold,[14] especially in the West. Even the terrorist attacks of '9/11', and then US Vice President Dick Cheney's 'one percent doctrine',[15] the fears of clashes between now nuclear-armed India and Pakistan, or the 2006 nuclear test by North Korea didn't really fundamentally alter this feeling of nuclear complacency. By the mid-2000s the nuclear disarmament movement was again back in full swing and nuclear reductions and nuclear peace (outside of the concern about North Korea and to a lesser extent Iran) became the norm. In the late 2000s organizations such as Global Zero even began to outline the steps towards complete nuclear disarmament in a relatively short time frame.[16] Terrorism would dominate the political debate in the early 2000s (reflected in films such as *The Sum of All Fears*—where a neo-Nazi faction attempts to detonate a nuclear bomb in Baltimore). But this would slowly be supplanted as the number one challenge facing humanity by the all-pervading logic of the 'climate crisis' in the mid-2000s. Outside of Delhi, Islamabad, or Seoul, nuclear weapons were hardly mentioned at all, and didn't appear as a major issue of public interest or concern.[17] In this way, the return to disarmament during this period can perhaps best be characterised as largely an elite project, and arguably driven as much by a perceived opportunity and ideology as it was by as any immediate security concern. Nevertheless, and while hindsight might suggest hubris, there really was a feeling that nuclear weapons were on the way out as large parts of the world appeared to become more peaceful, globalised, interconnected, and 'post-modern', and attention turned to the importance of what the prominent US

[14] Francis Fukuyama, *The end of history and the last man* (Washington, DC: Free Press, 1992).

[15] If the US believed that there was a 1% chance of a terrorist group using weapons of mass destruction (WMD) the US should act as if it was a certainty. See Ron Suskind, *The one percent doctrine: Deep inside America's pursuit of its enemies since 9/11* (London: Simon & Schuster, 2006).

[16] E.g. https://www.globalzero.org/reaching-zero/index.html. In 2012, former commander of US STRATCOM General James Cartwright suggested that a deployed force of 450 US nuclear weapons was possible. 'The retaliatory capability of 300 nuclear weapons is catastrophic', Centre for Arms Control and Non-Proliferation, 12 August 2012, https://armscontrolcenter.org/gen-cartwright-quotthe-retaliatory-capability-of-300-nuclear-weapons-is-catastrophicquot/.

[17] For a discussion of this, see Andrew Futter & Benjamin Zala, 'The return of nuclear great power politics (or why we stopped worrying about terrorists and the bomb)', *European Journal of International Security* 10:1 (2024), pp. 133–149.

scholar and sometime policy official Joseph Nye termed 'soft power'.[18] One commentator even went as far as to describe this period as a 'holiday from history'.[19]

But today this has changed, with the general public and policy elites having to relearn the grammar of nuclear risks, deterrence, military security, and hard power. For sure, those living in India, Pakistan, and parts of northeast Asia have been living with the reality of immediate and terrifying nuclear risks for over two decades, but this represented the exception rather than the general global experience. Perhaps reflective of this we have seen a return of nuclear culture to the Western mainstream, including in films such as *Oppenheimer*, the screening of the TV docu-drama *Chernobyl* and the rescreening of *Threads* in the UK, *House of Dynamite* on Netflix, and the release of Annie Jacobsen's book *Nuclear War*.[20] In a notably different way, nuclear weapons have gradually emerged as a source of cultural national pride in India, China, and elsewhere: the 2018 Bollywood film *Parmanu: The Story of Pokhran* and the 2019 Chinese drama *Ji Qing De Sui Yue*, or *The Passionate Years*, being good examples.[21] Whether the millions of 'computer geeks' playing *Fallout*, *Call of Duty*, or other apocalyptic video games contributes to this remains to be seen.[22]

This work is therefore intended as a wake-up call. There is nothing preordained about our nuclear future, but ignoring the changing world around us and hoping for the best is not a viable option either.

The book makes seven core arguments.

[18] Joseph S. Nye, 'Soft power', *Foreign Policy* 80 (1990), pp.153–71.

[19] The phrase is usually attributed to *The Washington Post* journalist George Will.

According to William Langewiesche, 'In America, the apparatus of nuclear war remained in place, but the subject came to occupy the sleepiest corners of the military complex. Paul Bracken lamented the change. He told me that after the Cold War, the smart money lay elsewhere'. In "The secret Pentagon war game that offers a stark warning for our times, *The New York Times Magazine* (2 December 2024), https://www.nytimes.com/2024/12/02/magazine/nuclear-strategy-proud-prophet.html

[20] Annie Jacobsen, *Nuclear war: A scenario* (London: Transworld publishers, 2024). Interestingly, both this book, and Jeffrey Lewis' earlier, *2020 commission on the North Korean nuclear attacks against the United States* (London: W.H. Allen, 2018), focus on North Korea as the main nuclear challenge. A slightly lesser known forerunner to this is Richard Shirreff, *War with Russia* (Columbus, OH: Coronet, 2016).

[21] See, for example, Souvik Kar, 'Mytho-nukes: The Indian popular and the nuclear debate', in Gitanjali Chawla & Sangeeta Mittal (eds.), *Indian popular fiction: Redefining the canon* (London: Routledge, 2022). A notable exception being 'Hava Aney Dey' ('Let the wind blow') released in 2004.

[22] See Cameron Hunter, 'The ambivalent nuclear politics of fallout video games', *Bulletin of the Atomic Scientists*, 17 October 2018, https://thebulletin.org/2018/10/the-ambivalent-nuclear-politics-of-fallout-video-games/; and Carolina Pantoliano, 'Playing doomsday: Video games and the politics of nuclear weapons', *Review of International Studies* (2025), pp. 1–21.

First, we have entered a new chapter of our nuclear story where the balance at the heart of what we think of as 'nuclear order' is being challenged. The nuclear world today looks very different to how it did a generation ago: it is more complex, contested, and capricious. This does not mean everything we thought or did in the past was wrong or has been superseded, that enduring problems have been solved, or that actors in different parts of the world share the same experience, but rather that we urgently need to rethink how we understand the nuclear condition for a changed context.

Second, the concept of nuclear ages is an imperfect but nevertheless useful construct to capture the nature, speed, and consequences of this transformation in the global nuclear order. However, the periodisation becomes more useful if its scope is expanded to look at our nuclear world more widely, and escapes the nuclear ethnocentrism that is common in nuclear studies (and which I hope to avoid here ...). It is important to reflect on the impact of how we choose to conceptualise our nuclear history, and the power dynamics that this reflects, as this fundamentally shapes the types of problems we emphasize and the solutions that might be considered as possible and practicable.

Third, technological advances—often captured under the 'Emerging and Disruptive Tech or EDT' moniker—are central to the Third Nuclear Age thesis, but the impact and implications of 'new' weapons, capabilities, and missions—and the beliefs and fears that they may generate—are far more nuanced than conventional wisdom often suggests. Many technological changes pointed to today can be traced back decades, and the rate of adoption of such technologies varies considerably across actors and regions. That said, technological changes, particularly the diffusion of strategic *non-nuclear* weapons—are challenging established notions of strategic stability, creating pressures for nuclear and non-nuclear proliferation, and are playing out within a new digitised, dispersed, real-time nuclear ecosystem.

Fourth, deterrence, assurance, and non-proliferation remain fundamental concepts in the Third Nuclear Age, but the different strategies being pursued to deter nuclear threats are becoming increasingly varied: there is no one-size-fits-all model. For a mixture of political, economic, and cultural reasons different actors are either doubling down on secure second-strike forces, developing or emphasising nuclear weapons for military applications on the battlefield, limited use and coercion, or pursuing a *non-nuclear* advanced conventional pathway to nuclear security. This in turn is making stability, predictability, and the future proliferation landscape potentially more complicated.

Fifth, we cannot properly understand our current nuclear world without taking a genuinely global approach that moves beyond the Western-centric literature and an overwhelming focus on nuclear weapons and the nuclear-armed states, and looks at different interests and objectives from non-nuclear-armed stakeholders too. Great power competition between the US and China, and the unfolding nuclear relationship between Washington, Beijing, and Moscow are central to the new nuclear era, but they are not the only games in town. A feeling that US power and perhaps also the liberal international order is weakening will challenge existing regional security complexes in Europe, the Middle East, and Asia that have long relied on the stability and predictability of a US-led global system, and could reopen questions of proliferation. At the same time, these dynamics may create opportunities for other actors to influence global nuclear politics, especially in the 'global south', by amplifying previously marginalised viewpoints and interests. We can think of this as the manifestation of nuclear globalisation.[23]

Sixth, the direction of travel in nuclear-arms control, diplomacy, activism, and scholarship in the Third Nuclear Age appears to be towards increased division, disagreement, and in some cases even demonisation. While nuclear order has always been contested, at almost every level consensus and collective action appear to be becoming more difficult to achieve. The bilateral arms control process that has seen massive reductions in nuclear stockpiles has stalled; the TPNW is a very vivid demonstration of the non-nuclear-armed world's frustration at the pace of disarmament via an incremental approach based on the NPT; and the battle continues across the activist, NGO, and academic ecosystem to challenge dominant paradigms and shape the construction of nuclear knowledge. This may be an inevitable result of a more fractured power structure within the global nuclear order—and possibly linked to the prominence of certain powerful individuals, or a reflection of a slow death of the Western liberal project that came to prominence a generation ago. But either way, it is not a sustainable way to build a peaceful future; we must find ways to talk, and envisage realistic nuclear and non-nuclear futures, rather than increasingly talk past each other.

Seventh, the risks associated with the continuing operation of nuclear technologies cannot be properly understood in isolation. Nuclear technologies are intrinsically linked with and co-constituted by the challenges posed by climate change and the quest for sustainable and equitable human development. While nuclear use has always had the potential to impact the

[23] I am grateful to Olamide Samuel for coming up with this term.

climate, today climatic effects may well impact nuclear operations, and most visions of our future as a species seem to demand more rather than less atomic energy generation. The unfolding geopolitical world therefore seems set to be one where civil nuclear application's experience a renaissance and will arguably spread much quicker than in the past, to both states and multinational companies. This in turn will reignite fears of accidents, radiological release, nuclear terrorism, and weapons proliferation. The Third Nuclear Age will therefore be coterminous with the widespread recognition of the challenges of the 'Anthropocene' as nuclear technologies diffuse, the climate crisis deepens, ever more people demand 'clean' energy, and fears return of nuclear impacts on our increasingly interconnected planet.

Siting above these not inconsiderable challenges, changes, and dynamics is a genuine concern about the level of elite political will, public engagement, and physical preparedness to respond to existential societal challenges. In the Third Nuclear Age this appears to be reducing rather than increasing. Putting these challenges in the 'too difficult box' is not a good way to go about securing humanity's future, neither is ignoring sincerely held but different views about how this future should be pursued.

This book is therefore intended as an attempt to both map out our complex and changing nuclear world and at the same time raise awareness of the dangers (and possibly opportunities) that nuclear technologies continue to pose to humanity. There is a fear that for the best part of a generation (the generation in which I grew up) many humans essentially forgot about the threats posed by nuclear weapons, or at best decided to ignore them. Sadly, this is no longer a viable approach (if indeed it ever was). Nuclear challenges are back at the centre of international politics, and this time there are more actors involved; more complex relationships and dynamics that could lead to nuclear flashpoints, accidents, and detonations; and perhaps less appreciation of the impact that nuclear use could have on us as a species.[24]

Whether or not one agrees with the framing of a Third Nuclear Age, most of us are living in a more dangerous nuclear world today than we have for at least the last 35 years and perhaps even longer. It is essential that we engage with the difficult nuclear challenges and risks and not simply hope that they go away.

[24] Notwithstanding the announcement of a UN panel on 'The effects of nuclear war in 2024', https://disarmament.unoda.org/panel-on-the-effects-of-nuclear-war/#:~:text=The%20Panel%20has%20been%20tasked,and%20ecosystems%2C%20in%20the%20days%2C.

There are therefore two main objectives of this work. First, to challenge and contribute to the existing debate on a New or Third Nuclear Age by arguing for a broader conceptual focus that looks beyond particular exotic technologies, the Washington-beltway view of great power competition, and the often-parochial focus upon maintaining the Western-led nuclear order, and at the same time offer a more nuanced understanding of change in nuclear politics. And second, to call for greater engagement, education, debate, and outreach when it comes to understanding nuclear issues. Taken together, the intention is to provide a grounded attempt to comprehend our evolving nuclear world, hopefully breathe life into public engagement with the nuclear phenomenon, and provide a basis for how we can think about our nuclear future.

Outline of the book

The book proceeds in six chapters, each focussing on a key component of the Third Nuclear Age. This structure is not meant to suggest linearity or causation—the chapters are about exploring the extent of change across different dynamics, mapping out what is and isn't new, how, and for whom. There isn't one nuclear smoking gun.

The main aim of Chapter 1 is to examine how we think about the periodisation of nuclear history, the significance of thinking in terms of three nuclear ages, and how this shapes our understanding of the nuclear order and what is or isn't possible. It makes the argument that while often geographically and culturally subjective—and reflective of the intellectual power and the hegemony of particular viewpoints—splitting nuclear history into different periods characterised by distinctive dominant trends and threats is a useful way to highlight continuity and change in nuclear politics, and to draw attention to the most pressing challenges at any given time. However, for a Third Nuclear Age framework to make analytical and intellectual sense, it is essential to move beyond the often-narrow confines of previous 'nuclear age' framings, take a global and holistic approach, and recognise that not everything changes in the same way or at the same time for all stakeholders.

Chapter 2 seeks to demystify the technological hype that underpins so much of the current fear at the heart of the Third Nuclear Age. Innovation in and diffusion of weapons and support systems as well as changes in the information environment over the past few decades are having an impact on all aspects of nuclear politics, stability, and operations, but the changes and implications are often more nuanced than the tendency for buzzwords and hyperbole reported in the media and other fora might suggest. AI, 'cyber', space weapons, and hypersonic missiles are all part of the Third Nuclear Age tapestry but tend to draw our attention away from the most consequential technological developments taking place. In reality, many questions and concerns posed by technology are either not that new or represent the latest iterations of established challenges, or their (possible) realisation may be some way off into the future. In this way, the most important development is what might be considered a 'quiet metamorphosis' as existing technologies and associated weapons systems that might impact global nuclear politics become smarter, real time, dispersed, commingled, and more complex.

Technological change is a key dynamic reshaping the ways, ends, and means of deterrence in the Third Nuclear Age, but as Chapter 3 explains this is playing out differently for different actors in different parts of the world. The most prominent example is the impact on those that already have nuclear weapons, with a more conspicuous split between states who describe themselves as 'responsible nuclear actors'—broadly united by a commitment to minimal nuclear forces and secure second-strike deterrence, and those increasingly labelled as 'irresponsible' due to an increasing interest in warfighting, battlefield, and coercive nuclear applications. The most transformative dynamic is potentially the group of states that have embraced a non-nuclear pathway as the primary means of deterring nuclear threats through advanced/strategic non-nuclear weapons—a strategy that simply would not have been viable in previous eras. Taken together this more prominent diversification in the methods used to achieve deterrence of nuclear threats, and even to fight wars against nuclear-armed opponents, is creating new challenges for stability, proliferation, and crisis management, and opening up new questions about the possibility of counterforce attacks.

Great power *nuclear* competition between the US, China, and Russia is undoubtedly central to the emerging nuclear context. But as Chapter 4 shows, this is only one dynamic that will impact the geopolitical shape and scope of the Third Nuclear Age. A gradual move away from relatively uncontested US hegemony towards a more multipolar nuclear world seems likely to fundamentally reshape nuclear security politics in Europe, south and northeast Asia, and possibly the Middle East, and create pressures for renewed proliferation or 'dominos'. It is also possible that middle powers and states from the 'global south' will play an increasingly important role in all aspects of global nuclear politics. Gradual shifts in global power balances may even progressively impact and challenge the historical Western dominance of the nuclear debate and institutions, which, in turn, may well give more prominence to issues central to non-nuclear-armed states, such the commitment to disarmament and access to nuclear technologies for development.

Chapter 5 examines how attempts to address the challenges posed by nuclear technologies are breaking down—and in some cases becoming more antagonistic—across different parts of the global nuclear order, and puts these in historical context. This starts with the breakdown and erosion of the nuclear-arms control architecture that was painstakingly built up over the First and Second Nuclear Ages, but which has now almost totally unravelled, and can be felt through increased disharmony in the international nuclear fora designed to address nuclear threats, through to a more combative clash between disarmers and deterrers, and different ontological visions of security across civil society and academia. The trend appears to be for two very different and sometimes hostile visions of the nuclear future to become increasingly entrenched with what appears to be diminishing prospects for meaningful dialogue or progress.

Finally, Chapter 6 makes the case for viewing nuclear challenges as just one component of a larger co-constitutive system of existential threats to humanity. In this way the Third Nuclear Age is intrinsically linked with the politics and challenges of the Anthropocene, namely the three-way relationship between nuclear technologies; equitable economic, societal, and civilisational development and progress; and the sustainability of our Manichaean relationship with the environment and the planet. Nuclear technologies are likely to spread in response to climate, scientific, and developmental demands; the changing climate will impact nuclear facilities and operations, and potentially create new security flashpoints; while the spectre of weaponisation, accidents, and attacks on nuclear infrastructure all remain acute and growing challenges. Ultimately, the (mis)use of nuclear technologies continues to represent the most severe and instantaneous threat to our planet, and the only (current) way to rapidly destroy ourselves as a species.

The Conclusion argues that we may be ill prepared either intellectually or societally to deal with the possibility of a major shock to the way that we live our lives. While this does not mean that everything in the Third Nuclear Age is new or novel—many entrenched and established dynamics and challenges remain—we do need to shift our frame of reference to the realities of what seems likely to be a more dangerous era. For sure, the Third Nuclear Age may see the amplification of a chorus of anti-nuclear voices, but the nuclear-armed states ultimately remain the most powerful in the international system, and this is likely to limit genuine change. In many ways, humans—especially outside of south Asia and the Korean peninsula—have learnt to live with and largely ignore nuclear risks since the start of the new millennium, but the future nuclear world will be one where preparing for a major nuclear disaster is a reality. Sadly, the majority of the global response to the recent COVID-19 pandemic gives some insight into the widespread lack of societal preparedness for a major nuclear disaster, prospects for global cooperation, and why returning to thoughts of civil defence is complicated. Likewise, the perception that societies are becoming less engaged, aware, or even interested in nuclear issues creates problems for 'raising the nuclear IQ'. None of this seems to bode well for our nuclear future ...

But ultimately, the nuclear challenge is humanmade—there is nothing pre-ordained—so in theory there is nothing to prevent us finding a way to manage and possibly even transform our current predicament. As the British scholar Nick Ritchie recently asked a group that I was part of: might humans in 500 years' time look back on the nuclear era as an aberration ...?

1

Whose nuclear age is it anyway?

Introduction: Constructing nuclear knowledge

In a speech at the Royal United Services Institute (RUSI) in London in late 2024, then chief of the United Kingdom (UK) Defence Staff, Admiral Tony Radakin declared that the world had moved into a 'Third Nuclear Age'.[1] As he explained:

> Global power is shifting, and a third nuclear age is upon us. The era of state competition primarily through geo-economics has shifted to a resurgence of geo- politics. And it will last decades ... [the Third Nuclear Age] is defined by multiple and concurrent dilemmas, proliferating nuclear and disruptive technologies, and the almost total absence of the security architectures that went before.

Admiral Radakin's speech reflected a growing use of the idea of a Third Nuclear Age in the media, by academics, and by different officials both in the UK and beyond. But what does it mean to be in a 'Third' Nuclear Age? And how might this be different from a 'First' and 'Second' Nuclear Age that had presumably preceded it?

The concept of *the* nuclear age and nuclear *ages* has become commonplace in the nuclear studies lexicon. But these are far from exact terms, often understood in different ways by different actors.[2] Did the nuclear age begin underneath a football stand in December 1942 when Enrico Fermi's 'pile' went critical? In July 1945 with the Trinity Test or a few weeks later with the bombings of Hiroshima and Nagasaki? Or did it begin decades earlier

[1] 'Chief of the Defence Staff warns of a "Third Nuclear Age" in RUSI annual lecture', 4 December 2024, https://rusi.org/news-and-comment/rusi-news/chief-defence-staff-warns-third-nuclear-age-rusi-annual-lecture.

[2] As the Carnegie Endowment's Ariel Levite put it: 'assigning chronological cut-off points between the "nuclear ages" inevitably involves some arbitrary decisions which might make more than a few historians and political scientists uncomfortable.' Ariel E. Levite, 'Heading for the fourth nuclear age', *IFRI Proliferation Papers*, Winter 2009, p. 10.

The Third Nuclear Age. Andrew Futter, Oxford University Press. © Andrew Futter (2026).
DOI: 10.1093/9780198944126.003.0002

with Marie Curie's discovery of radium in 1898, Albert Einstein's famous equation $e=mc^2$ in 1905, Ernest Rutherford's discovery of the atomic nucleus in 1911,[3] or the first theories of nuclear decay that led to influential sci-fi novels such as H.G. Wells' *The World Set Free* published in 1914?[4] As early as the 1920s, Winston Churchill was writing about 'tremendous and awful powers ... upon a scale that can annihilate whole nations'.[5] But despite the fact that our knowledge of nuclear technology grew exponentially over the first four and a half decades of the 20th century, most standard Western accounts of nuclear history start in August 1945.

Most of us alive today have lived the majority of our lives surrounded by the reality of nuclear technologies. But should we treat this era as a single period defined principally by the ability to split the atom, or does it make sense to divide this history up into different sections, with different characteristics and focal points? Is this a useful way to distinguish between our nuclear past, what is happening today, and what might happen in the future? And if so, what should this be based upon? And whose framings and experiences of this history matter? How we think about and conceptualise our nuclear history is important, not least because it fundamentally structures how we understand and construct the gravest risks and opportunities of the nuclear world around us, whose voices are amplified or excluded, and what might or might not be deemed acceptable and possible.

The standard Western periodisation of our atomic history is that there have been two 'nuclear ages'.[6] A First Nuclear Age is said to have begun in 1945 and is broadly contiguous with the Cold War rivalry between East and West that dominated international politics for a generation. A Second Nuclear Age is said to have emerged in the 1990s as the era of United

[3] For a fascinating overview of early nuclear history, see Frank Close, *Destroyer of worlds: The deep history of the nuclear age 1895–1965* (London: Allen Lane, 2025).

[4] H.G. Wells, *The world set free* (London: Macmillan & Co, 1914).

As Wells put it, 'It could mean a change in human conditions that I can only compare to the discovery of fire, that first discovery that lifted man above the brute. We stand to-day towards radio-activity as our ancestor stood towards fire before he had learnt to make it. He knew it then only as a strange thing utterly beyond his control, a flare on the crest of a volcano, a red destruction that poured through the forest. So it is that we know radio-activity to-day. This-this is the dawn of a new day in human living. At the climax of that civilisation which had its beginning in the hammered flint and the fire-stick of the savage, just when it is becoming apparent that our ever-increasing needs cannot be borne indefinitely by our present sources of energy, we discover suddenly the possibility of an entirely new civilisation'. p. 23.

[5] Graham Farmelo, *Churchill's bomb: A hidden story of science, war, and politics* (London: Faber, 2013), p. 43.

[6] The idea of a 'First Nuclear Age' only appeared 'post-hoc' once scholars and policymakers started speaking of a Second Nuclear Age in the 1990s.

States (US) unipolarity shifted nuclear concerns towards horizontal proliferation to 'rogue states' and other 'undeterrable' actors. The Cuban Missile Crisis of 1962 arguably captures the fears at the heart of the First Nuclear Age, while the anxieties caused by 'irrational' leaders of aspirant nuclear-armed 'rogue' states and terrorists armed with the bomb reflect the canon of thought in the Second Nuclear Age. Both would become central to the overarching narratives of 'MAD (mutual assured destruction) and deterrence stability' in the First Nuclear Age and the priority of 'nuclear security and counterproliferation' in the Second Nuclear Age.

While this periodisation is instructive, it nevertheless reflects a particular framing of nuclear history that fits with and is shaped by the dominant political and normative actor(s) in the international system. What is therefore needed is a broader, more reflective conceptualisation of what matters and to whom in the global nuclear order, the power structures this reflects, and a framework that engages with voices, viewpoints, and challenges both inside and outside the mainstream. The world is moving into a Third Nuclear Age, as Admiral Radakin and others have suggested, but how we think about and analyse this new era needs to avoid, or at least acknowledge, the cultural nuclear ethnocentrism and intellectual power structures often intrinsic in First and Second Nuclear Age framings.

Nuclear history is written by the victors

How we choose to label and periodise history—including nuclear history—is not a perfect science, is often inherently subjective, and can be the product of dominant cultural or social norms and groups, and possibly even used instrumentally.[7] Periodisations, such as 'La Belle Époque', the 'Enlightenment', the 'Industrial Revolution', the 'Ming Dynasty', the 'swinging 1960s', and the 'Ice Age', all provide framings that differ markedly in terms of time period, key actors, referent object, and geopolitical scope. They are also periodisations that have often been imposed post hoc by particular cultural groups, and only really make sense when compared with what followed.

[7] For a good discussion of this, see Almuth Ebke & Christoph Haack, 'Periodisation and modernity: An introduction', *History of European Ideas* 51:2 (2024), pp. 307–20; Reinhart Koselleck, *Futures past: On the semantics of historical time* (Cambridge, MA: MIT Press, 1985); Daniel Woolf, A. Mahler, & C. Zwierlein, 'Historical periodization: An exploration and defence', in Cornel Zwierlein & Andreas Mahler (eds.), *Zeiten bezeichnen. Frühneuzeitliche Epochenbegriffe. Europäische Geschichte und globale Gegenwart / Labelling times: The 'early modern'. European past and global now* (Wiesbaden: Harrassowitz, 2023), pp. 29–56.

How we choose to think about and divide up our *nuclear* history is therefore not a neutral or dispassionate exercise, and this means we have to think carefully about whose history is being recorded and how. As Thomas Donnelly and David Trachtenberg put it in a piece for the *American Enterprise Institute*:

> Be they social scientists or political pundits, academics and analysts are forever dividing history into epochs and eras that too often impart a false sense of clarity to the muddy course of human events.[8]

There is no one definitive account of nuclear history or one model that all agree upon to understand changes in our nuclear world over time. Different dynamics and moments are important for different actors for different reasons and shape shared historical experiences and narratives. Key moments of change or demarcation for one particular culture or group, or in one part of the world, are not necessarily universally shared or reflect the experiences in different parts of society or different parts of the world. Attempts to categorise nuclear history into particular periods is therefore always somewhat arbitrary and reflective of the writers' personal preferences, geographical location, and often political power structures. It is also often reflective of the dominance of particular ideas, cultures, and theories at any given time, and the ability of some groups to 'shape the agenda' more than others.[9] But this does not mean that trying to impose a certain order on nuclear history is impossible, rather that any such efforts must take these nuances into account.

A growing body of scholarly work suggests that the history of nuclear technology—and the accompanying academic literature—has been heavily shaped by power—whether political, military, economic, or ideological—and by the interests and agendas of the most powerful actors. French philosopher Michel Foucault went as far as to argue that this power has been enforced by the diffusion of 'discourses, norms, institutions, and practices'.[10] The argument continues that these power dynamics often obscure certain aspects of nuclear history, such as the humanitarian costs, environmental impacts, and the voices of particular groups, and that this subjectivity has

[8] Thomas Donnelly & David Trachtenberg, 'Toward a new "new look": US nuclear strategy and forces in the third atomic age', American Enterprise Institute, 2010, p. 1.

[9] Indeed, in this way it maps onto what political scientist Steven Lukes described in the 1970s as 'agenda shaping power'. See Steven Lukes, *Power: A radical view* (London: MacMillan, 1974).

[10] See Michel Foucault, *Power/knowledge: Selected interviews and other writings, 1972–1977* (New York: Knopf Doubleday Publishing Group, 1980).

led to the marginalisation of different experiences or events that don't fit neatly into the established paradigms. At the same time, these dynamics allow the most powerful actors to shape the nuclear story and the nature of nuclear knowledge. Therefore, it is essential to critically examine how nuclear history is constructed and whose interests are being served in these narratives.[11]

A conflation of factors—namely the emergence of the US as the most powerful actor in global politics after 1945, the vast resources brought to bear on the nuclear 'problem' by the Pentagon, wealthy universities and research institutes such as the RAND Corporation, and an overwhelming concentration of research and writing on nuclear issues in the US and to a lesser extent western Europe, has created an Anglo-centric, English-language canon of scholarly and professional thought on nuclear history. This is not automatically a bad thing, but must be understood as reflecting just one interpretation of our atomic past—albeit one that has been very powerful in shaping global framings, perceptions, and norms.

The idea that nuclear history can be split into two periods or 'ages' either side of the Cold War is a direct product of this context and genesis, and reflects an overwhelmingly Washington-based view of the nuclear world, and especially the major nuclear challenges facing the US at two particular historical moments. For large parts of the world, the threats posed by nuclear weapons and technologies have remained fairly constant over time, or at best didn't shift markedly with the end of the Cold War. A Second Nuclear Age may have begun for the US and its allies after the collapse of the Soviet Union, but this does not automatically make it a globally shared nuclear inflection point. One could argue for instance that a First Nuclear Age began in the Middle East in the 1960s and south Asia in the 1970s and that these regions are still effectively in their First Nuclear Age given that the core nuclear security dynamics have changed little since this time.[12]

Nevertheless, the concept of *the* nuclear age[13] and nuclear *ages* has become commonplace in the security studies and international relations lexicon. Nuclear ages are a valuable, albeit very simplified, way to break

[11] For a discussion on this concept, see Målfrid Braut-Hegghammer, 'Proliferating bias? American political science, nuclear weapons, and global security', *Journal of Global Security Studies* 4:3 (2019), pp. 384–92. See also Nick Ritchie, 'A hegemonic nuclear order: Understanding the Ban Treaty and the power politics of nuclear weapons', *Contemporary Security Policy* 40:4 (2019), pp. 409–34.

[12] As a participant in a workshop that I organised in New Delhi in early 2025 told me—'for India the first nuclear age began the day that China tested its first nuclear device in 1964'.

[13] One contemporary example being, Serhii Plokhy, *The nuclear age* (London: Allen Lane, 2025).

up nuclear history, but also a framing that is clearly based on a particular experience and viewpoint—principally that of US elites, policymakers, and strategists. That said, the construct of a Second Nuclear Age as a dominant ordering mechanism and point of departure is treated unproblematically in the vast majority of nuclear literature and is used in a way that assumes a certain meaning: the only real exceptions being several paragraphs in the books by Colin Gray and Paul Bracken.[14] As Gray explained;

> Historical periodisation can be apparently arbitrary, is often contestable even on its own leading criterion, and is neither right nor wrong. In common with definitions, periodisation is either more or less useful and, on the evidence available, more or less plausible. Letters of fire in the sky did not proclaim on 6 August 1945 that the (first) atomic, or nuclear, age had dawned. Similarly, on 22 December 1989—the day that the Berlin Wall began to be dismantled—no unarguably authoritative source insisted that a second nuclear age was dawning. Although the concepts of first and second nuclear ages are intellectual constructs, these concepts are both massively supportable empirically and, above all else, they are useful.

Key to the nuclear ages concept is what scholars, professionals, officials, and the wider public should focus on. The dominant theme prioritises a mixture of international structure and overarching nuclear threats and how these might be deterred, and in this way is inherently 'realist'.[15] A different focus could easily break down these periods further, Ariel Levites' four nuclear ages, for example,[16] or something closer to the seven historical periods looked at by William Walker as part of his examination of the evolution of nuclear 'order'.[17] Other dominant themes have been to split nuclear history into smaller periods based on US presidential administrations,[18] or

[14] Colin Gray, *Second nuclear age* (Lynne Rienner Publishers, Boulder Colorado, 1999), pp. 5–9; Paul Bracken, *Second nuclear age* (St Martin's Griffin, New York, 2013), pp. 1–14.

[15] As in the school of thought in International Relations scholarship. For an excellent contemporary exposition of this field, see Patrick Porter, *How to survive a hostile world: Power, politics and the case for realism* (Stanford, CA: Stanford University Press, 2025).

[16] Levite, 'Heading for the fourth nuclear age'. Levite's model broke nuclear history into: First Nuclear Age, 1945–67 (dealing with the appearance of nuclear weapons); Second Nuclear Age, 1968–92 (Non-Proliferation Treaty (NPT) onwards: building and managing stability); Third Nuclear Age, 1993–2010 (characterised by 'complacency and disillusionment'); Fourth Nuclear Age 2011– (with several key dynamics including proliferation, crisis, or nuclear terrorism).

[17] William Walker, *The perpetual menace: Nuclear weapons and international order* (London: Routledge, 2012).

[18] A good example being John Lewis Gaddis', *Strategies of containment* (Oxford: Oxford University Press, 2005).

as Lawrence Freedman's—more recently with Jeffrey Michaels—magisterial *Evolution of Nuclear Strategy* by looking at thematic dynamics in a broadly historical narrative.[19] Other alternative framings might be akin to Lawrence Wittner's three phases of nuclear disarmament,[20] a focus on nuclear weapons and the Anthropocene (dealt with later in this book ...), or a perspective drawn from postcolonial and decolonial scholarship.

For a vast majority of policymaking elites across the world, the worries and opportunities posed by nuclear technologies has changed little over time, or the key moments of change are different to those experienced by others. For example, while nuclear proliferation concerns came to dominate Western views of the Second Nuclear Age, for many others, the 1990s and early 2000s were a time of great optimism for nuclear disarmament (as a process, if not necessarily as an end state). A different narrative might therefore illuminate the fact that while the end of the Cold War was clearly a momentous global event, it didn't necessarily change the calculations and viewpoints of most states within the global system. If anything, it reinforced the hope that meaningful progress could be made towards nuclear abolition, as evidenced in 1995 when the Non-Proliferation Treaty (NPT; it had originally been given a 25-year lifespan) was extended indefinitely, which, it was hoped, would lay the foundations for a more peaceful world; the South African decision to unilaterally disarm; and by the nuclear-weapons-free zones that were agreed in Southeast Asia (1995), central Asia (2009), and Africa (2009).

Neither does the 1990/1 cut-off point really reflect the reality of the shift towards different nuclear challenges: all of the nuclear proliferation 'threats' that would come to animate the Second Nuclear Age discourse began—and in some cases materialised—in the First Nuclear Age.[21] As leading chronicler of the Second Nuclear Age thesis, Paul Bracken of Yale University put it,

> The tendency is to think of victory in the Cold War as the turning point of the modern era. But that this not how it looks to much of the worlds people. For them, it's a closing of a five-hundred-year period of Western domination, not fifty years of Cold War. Looking at the second nuclear age in terms of the Cold War may well be a Western conceit.[22]

[19] Lawrence Freedman & Jeffrey Michaels, *The evolution of nuclear strategy*, 4th ed. (London: Palgrave MacMillan, 2019).
[20] Lawrence Wittner, *Confronting the bomb: A short history of the world nuclear disarmament movement* (Stanford, CA: Stanford University Press, 2009).
[21] India and Pakistan probably already had a nuclear weapons capability, Iraq and North Korea, and maybe others had advanced programmes by the early 1990s.
[22] Bracken, *The second nuclear age*, p. 122.

Ultimately, the nuclear ages literature and concept reflect particular constructs and beliefs, not necessarily global ones, largely because of the dominant role of the US in the international system and influential global fora during this time. The key to understanding this transition is therefore in how influential actors in the US 'spoke and acted' this into reality, which in turn was reified by the fact that leading American institutions were the centre of gravity of nuclear studies scholarship, and a smallish group of intellectuals dominated the production of nuclear knowledge during this time.[23]

The nuclear ages framing has so far focussed largely on those actors possessing or seeking to possess nuclear weapons and how they might be deterred. And specifically focussed on those actors that might pose a nuclear threat to the US, Western society and way of life, and more recently 'the liberal international order'. One could, for example, look through the lens of policymakers in Moscow, Beijing, New Delhi, and elsewhere and see not only a very different picture of the nuclear landscape in the 1990s and 2000s and a very different perception of the major nuclear threat, but also perhaps far less importance placed on 1990/1 as the major demarcation point.[24]

But this is not to say that splitting nuclear history up into separate 'ages' or periods is not useful—it is and can help us to understand what is motivating different actors at different chronological periods, especially the most powerful. In particular, it is a helpful way to capture the balance between continuity and discontinuity in the global nuclear order over time: the shift between nuclear ages representing moments or periods where trends of discontinuity or even 'rupture' have dominated. This means that despite the inherent US/Western ethnocentrism, the nuclear ages framing can be—and is—useful. The value of splitting nuclear history in this way is that it provides focus and gives clarity as to what the major nuclear threats, challenges, and dangers facing the global nuclear order were—or are—at any one point in the story. It also, importantly, allows us to see how these change over time, and to begin to think about how they came about. But as is discussed later in this chapter, the key to maximising the usefulness of the nuclear ages framing is to broaden out the scope of what constitutes the challenges and

[23] I am grateful to Ben Zala for this thought/observation, while discussing an earlier version of this chapter.

[24] For example, in an interview with the author in 2023, Russian expert Dmitri Stefanovich remarked that: 'The Western idea of two nuclear ages has some traction in Russia, but Russia has a different conceptualisation of this time period: up to the 1970s and then the first era of arms control; then the non-nuclear challenges of the 1980s; the reliance on nukes in 1990s due to conventional weakness; followed by ABMT withdraw in 2002 as significant in the transformation of Russian thinking, then a new age starting in the mid-2010s'.

opportunities of a Third Nuclear Age, and engaging more directly with the nature of change across different aspects of the global nuclear order. And, as is explained below, this framework indicates that there is ample evidence to suggest that we have moved beyond the Second Nuclear Age construct.

The concepts of a First and Second Nuclear Age

The concept of a First Nuclear Age is something that has been retrospectively applied to the period of global history that began in 1945 and lasted up until 1990. As far as I am aware, none of the major policymakers, commentators, or scholars working or writing in this period referred to or thought of themselves as living in or writing about a specific 'First' Nuclear Age. The temporalisation comes because of a feeling that the 1990s represented a major break from the past, and that a new and different era of global *and* nuclear politics had come to pass. The First Nuclear Age can therefore only be understood by comparison with what came after—the Second Nuclear Age.

The prevailing modelling of nuclear ages (or at least the canon of literature) suggests that between 1945 and 1990, the major concern within the global nuclear order was avoiding major interstate war between the nuclear-armed superpowers, the risks posed by nuclear arms racing, and to a slightly lesser extent preventing horizontal nuclear proliferation to 'middle powers' (both friends and foes). Consequently, and as is discussed in more detail in Chapter 5, the academic fields of security studies and international relations were largely dominated by rationalist and realist thinking about deterrence, arms control, and strategic stability. Notwithstanding the agreement of the NPT in 1968 (committing the five then nuclear-armed states to *work towards* disarmament in return for non-proliferation and access to nuclear technology for all), the emergence of global activist groups against the bomb, and the establishment of nuclear-weapons-free zones across large parts of the globe, disarmament as an end goal largely remained in the background.

The core tenets of what would become the First Nuclear Age construct were that this was a period characterised by nuclear bipolarity and dominated by the nuclear relationship between Washington and Moscow. For sure, there were other nuclear-armed actors during this period, beginning with the UK as early as 1952, but these were largely the supporting cast in the central Cold War drama. Major concerns were the nuclear arms race and the threat of a massive nuclear exchange between the superpowers that might have planetary-scale implications. Accordingly, First Nuclear Age literature focussed on managing these twin problems through mechanisms such as deterrence, extended deterrence, strategic stability, and arms control. The period gave rise to nuclear (and often overtly masculine[25]) epithets such as MAD, secure second-strike capabilities, throw weights, 'hardness of targets', and 'penetration aids', as well as institutions such as the International Atomic Energy Agency and the NPT designed to prevent future nuclear proliferation. This is not to say that the First Nuclear Age was somehow homogenous, the literature on 'three waves of deterrence theory' across the 1950s, 1960s, and 1970s captured by the distinguished American political scientist Robert Jervis[26] is just one example of how things had changed, but that the central challenges and structure remained fairly constant in the mainstream debate.

Much of what would become First Nuclear Age literature was bound up with trying to understand the 'nuclear revolution' and attempting to temper it.[27] As early as 1946, US scholar Bernard Brodie had captured the essence of the new types of challenge posed by nuclear weapons in his groundbreaking edited book *The Ultimate Weapon* and set in train what would become the dominant paradigm of nuclear deterrence. As Brodie memorably put it: 'Thus far the chief purpose of our military establishment has been to win wars. From now on its chief purpose must be to avert them. It can have almost no other useful purpose'.[28] Much of what followed would be directly linked to managing and understanding 'deterrence', such as Albert Wohlstetter's, 'The Delicate Balance of Terror' in 1959,[29] Brodie's, *Strategy in the Missile Age* 1959,[30] and Thomas Schelling's, *Strategy of Conflict* in 1960.[31] In Europe, this thinking was to some extent mirrored by the French philosopher and political thinker Raymond Aron in the 1950s and 1960s.[32] In 1961, the Australian academic Headley Bull published *The Control of the Arms Race* which marked the beginning of thinking about arms control, but still

[25] As Carol Cohn might say, highly gendered 'technostrategic' language. See Carol Cohn, 'Sex and death in the rational world of defense intellectuals', *Signs* 12:4 (1987), pp. 687–718.

[26] Robert Jervis, 'Deterrence theory revisited', *World Politics* 31:2 (1979), pp. 289–324.

[27] Michael Mandelbaum, *The nuclear revolution* (Cambridge: Cambridge University Press, 1981).

[28] Bernard Brodie, *The absolute weapon: Atomic power and world order* (North Stratford, NH: Ayer Company Publishers, 1946).

[29] Albert Wohlstetter, 'The delicate balance of terror', *Foreign Affairs* 37:2 (1959), pp. 211–34.

[30] Bernard Brodie, *Strategy in the missile age* (1959).

[31] Thomas Schelling, *Strategy of conflict* (Cambridge, MA: Harvard University Press, 1960).

[32] See Barry Cooper, 'Raymond Aron and nuclear war', *Journal of Classical Sociology* 11:2 (2011), pp. 203–24.

broadly within the existing intellectual paradigm.[33] All of this was later captured expertly by the American author/historian/journalist Fred Kaplan in his masterful chronicle *The Wizards of Armageddon*.[34]

These initial attempts to wrestle with the nuclear condition would be followed two decades later by articles written by leading scholars debating the feasibility of 'winning' nuclear wars, notably the work of Herman Kahn,[35] Colin Gray, and Keith Payne, but still very much within the bipolar nuclear superpower construct.[36] Towards the end of this period, Robert Jervis', *The Meaning of the Nuclear Revolution* (published in 1989) captured the essence of First Nuclear Age thinking by arguing that nuclear weapons represented a shift in the logic of international relations.[37]

With a few notable exceptions, such as the British polymath and Nobel-prize-winning physicist P.J. Blackett,[38] the philosophers John Paul Sartre[39] and Bertrand Russell,[40] pioneering feminist writers such as Carol Cohn,[41] and Carl Sagan and others' warnings about 'nuclear winter',[42] First Nuclear Age academic literature largely mirrored the dominance of traditional Strategic Studies and rationalist and realist thinking about deterrence, arms control, and strategic stability.[43]

The argument continues that after 1991 and the dissolution of the Soviet Union, global nuclear politics and global nuclear order changed—at least from the perspective of US military and policy elites.[44] Fears of proliferation to 'regional powers', and particularly to 'rogue' actors that might gain

[33] Headley Bull, *The Control of the arms race* (New York: Praeger, 1961).
[34] Fred Kaplan, *The wizards of Armageddon* (New York: Simon & Schuster, 1984).
[35] Herman Kahn, *Thinking about the unthinkable in the 1980s* (New York: Simon and Schuster, 1984).
[36] E.g. Colin Gray & Keith Payne, 'Victory is possible', *Foreign Policy* 39 (1980), pp. 14–27.
[37] Robert Jervis, *The meaning of the nuclear revolution* (Ithaca, NY: Cornell University Press, 1989).
[38] E.g. P.M.S. Blackett, *Fear, war, and the bomb: Military and political consequences of atomic energy* (Columbus, OH: McGraw-Hill, 1949).
[39] Ben Zala, '"No one around to shut the dead eyes of the human race": Sartre, Aron, and the limits of existentialism in the Nuclear Age', *Review of International Studies* 49:5 (2023), pp. 795–812.
[40] Bertrand Russell, *Common sense and nuclear warfare* (Hemel Hempstead: George Allen & Unwin Ltd, 1959).
[41] Cohn, 'Sex and death in the rational world'.
[42] Carl Sagan, 'Nuclear war and climatic catastrophe: Some policy implications', *Foreign Affairs* 62:2 (1983/4), pp. 257–292.
[43] This is not meant to be comprehensive—intellectual challenges to the dominant paradigm are explored in Chapter 5.
[44] Keith Payne has argued that 'many of the dominant features of the second nuclear age began to emerge by the 1980s ...'. Keith Payne, *Deterrence in the second nuclear age* (Lexington, Kentucky: The University Press of Kentucky.1996), p. 9.

access to unsecured nuclear material and facilities, especially after the terrorist attacks of 11 September 2001, came to dominate the Western debate. The main concern in Washington was no longer great power nuclear war between the heavily nuclear-armed East and West, but preventing various 'irrational' actors from getting and possibly using the bomb. This would increasingly be achieved through 'a new triad of offensive and defensive systems', as well as through strengthened non-proliferation and counter-proliferation mechanisms. This would become formalised as part of what would become known as the 'The Bush Doctrine', and codified in the 2002 US National Security Strategy and 2002 US Nuclear Posture Review.[45] The idea of disarmament returned in the 2000s—centred on an article published by four prominent US statesmen in 2007[46] and the speech by then US President Barack Obama in Prague in 2009[47]—and significant reductions were made to global nuclear stockpiles, but ultimately nuclear abolition stalled. Ironically, the Second Nuclear Age probably began its slow death after the *successful* 2010 NPT review conference and the much-celebrated 'US-Russian reset'—indelibly inked in the memory by a picture of US Secretary of State Hilary Clinton and Russian Foreign Minister Sergei Lavrov smiling while holding a big red button.[48]

With the exception of an article by the American historian Daniel Yergin in 1977,[49] literature with the label of a Second Nuclear Age did not appear immediately. American defence expert and former US government official Fred Iklé wrote in 1996 that 'with the Cold War now over, the drama of the nuclear age has returned for a second act';[50] former advisor to the US secretary of defence, Keith Payne published *Deterrence in the Second Nuclear*

[45] George W. Bush, 'The National Security Strategy', White House Archives, September 2002, https://georgewbush-whitehouse.archives.gov/nsc/nss/2002/. The 2002 NPR remains classified, but the foreword is available here: https://sgp.fas.org/news/2002/01/npr-foreword.pdf.

[46] Henry Kissinger, George Schultz, William Perry, & Sam Nunn, 'A world free of nuclear weapons', *Wall Street Journal*, 4 January 2007.

[47] 'Remarks by President Barack Obama in Prague as delivered', The White House Office of the Press Secretary, 5 April 2009, https://obamawhitehouse.archives.gov/the-press-office/remarks-president-barack-obama-prague-delivered.

[48] 'US-Russia relations: "Reset"', 24 June 2010, https://obamawhitehouse.archives.gov/the-press-office/us-russia-relations-reset-fact-sheet.

[49] Daniel Yergin, 'The terrifying prospect: Atomic bombs everywhere', *The Atlantic*, April 1977, https://cdn.theatlantic.com/media/archives/1977/04/239-4/132568249.pdf.

Yergin associated the move into a Second Nuclear Age with the Indian 'peaceful' test of 1974 and the proliferation cascade that he believed this foretold.

[50] Fred Charles Iklé, 'The second coming of the nuclear age', *Foreign Affairs* 75:1 (1996), p. 126.

Age in 1996; Colin Gray published his *Second Nuclear Age* book in 1999, and Paul Bracken published a book and article in 1999 and 2000 outlining what he saw as the new era in nuclear politics.[51] By the time of the terrorist attacks in the US on 11 September 2001, the idea of the move into a new and different period in nuclear history had begun to take hold, at least in the US and in a handful of other Western policy communities.

The core tenets of the Second Nuclear Age were that the nuclear threat no longer came from a heavily armed superpower (Soviet Union / Russia) but from smaller, regional powers seeking to possess weapons of mass destruction (WMD). 'We have slain a large dragon, but we live now in a jungle filled with a bewildering variety of poisonous snakes. And in many ways, the dragon was easier to keep track of' as the much-quoted remark from CIA Director James Woolsey's testimony to the 1993 Senate Confirmation Hearing had it.[52] But this did not necessarily mean that these challenges began after 1990, but that they had now risen to prominence as part of the new geopolitical milieu. As Paul Bracken explained: 'My definition of the second nuclear age—the spread of the bomb for reasons that have nothing to do with the Cold War—allows for countries to have gotten the bomb during the Cold War for reasons that had little to do with that conflict'.[53]

The main theme inside the Washington beltway was that the major threat to US security now came from proliferation to actors that may not act in the same way, or with a post hoc notion of rationality, as the Soviet Union had done before. As Keith Payne cautioned,

> the proliferation of weapons of mass destruction, advanced conventional weapons, and modern means of delivery (for example, ballistic and cruise missiles) appear to be promising otherwise third-rate military powers access to weapons of unprecedented lethality.[54]

Accordingly, (US) policy responses shifted from a reliance on retaliatory nuclear deterrence, to options that allowed for deterrence by denial, including measures to defeat limited missile threats through active defences and by

[51] Paul Bracken, 'The Second Nuclear Age', *Foreign Affairs* 79 (2000), pp. 146–57.

[52] 'Nomination of R. James Woolsey', Hearing before the Select Committee on Intelligence of the United States Senate, Tuesday 2 February and Wednesday 3 February 1993, p. 77, https://www.intelligence.senate.gov/wp-content/uploads/2024/08/sites-default-files-hearings-103296.pdf.

[53] Bracken, *The second nuclear age*, p. 96.

[54] Payne, *Deterrence in the second nuclear age*, p. 9.

conducting precision missile strikes against nuclear targets. This also linked in with a powerful idea that a 'revolution' was underway in military affairs,[55] and that a 'fourth wave' of deterrence theory focussed on deterring WMD terrorism was shaping the literature.[56]

The non-proliferation 'leg' of the NPT also took on a new importance, at least for those in the West, after the treaty was extended in 1995. UK-based academics, Campbell Craig and Jan Ruzicka, even went as far as describing 'Non-proliferation as part of the [US] hegemonic stability project'.[57] While others, such as French political scientist Benoit Pelopidas and colleagues argued that the post-Cold War framing of nuclear order in this way created certain path dependencies and reduced the scope for broader debate.[58]

The challenge at the heart of this construct was actually twofold: comprising both those states that represented a direct threat to the US/West and those actors that were proliferating for other reasons. In this way the concept of a Second Nuclear Age became entwined with another powerful geopolitical framing—that of the 'Global War on Terror'.[59] The Second Nuclear Age focus on 'rogue' states and non-state actors representing the major nuclear security threat also dovetailed with a concerted effort to 'make the world safe for democracy' and mirrored the quest for a 'liberal international order' underpinned by economic globalisation. After the nuclear war scares of the early 1980s,[60] this period of 'hegemonic stability' appeared to suggest that nuclear risks would only occur at the margins. The doyen of game theory, Thomas Schelling, seemed to capture this moment perfectly by describing it as a period of 'nuclear quiet' in a 2009 essay.[61]

[55] See, for example, Eliot A. Cohen, 'A revolution in warfare', *Foreign Affairs* 75:2 (1996), pp. 37–54.

[56] Jeffrey W. Knopf, 'The fourth wave of deterrence research', *Contemporary Security Policy* 31:1 (2010), pp. 1–33.

[57] Campbell Craig & Jan Ruzicka, 'The non-proliferation complex', *Ethics and International Affairs* 27:3 (2013), p. 336.

In a later email exchange with the author, Jan Ruzicka remarked that 'I would go as far as to say that the rise of the NPC is inextricably linked to the idea of the second nuclear age. In other words, no second nuclear age without the rise of the NPC and vice versa.'

[58] Benoit Pelopidas, Hebatalla Taha, & Tom Vaughan, 'How dawn turned into dusk: Scoping and closing possible nuclear futures after the Cold War', *Journal of Strategic Studies* (2024), pp. 1–23.

[59] See Andrew Futter & Benjamin Zala, 'The return of nuclear great power politics (or why we stopped worrying about terrorists and the bomb)', *European Journal of International Security* (2024).

[60] Particularly in 1983. See, for example, Taylor Downing, *1983: The world at the brink* (London: Little, Brown, 2018).

[61] The full quote reads: 'Except for some "rogue threats", there is little that could disturb those quiet nuclear relations among the recognized nuclear nations. Thus, nuclear quiet should not be

A few years later, Paul Bracken's 2013 *Second Nuclear Age* book also brought out an attendant dynamic, that another component of the new nuclear context was a shift in the locus of global (nuclear) politics to 'the East'. Again, mirroring a wider trend in largely Western literature that a power transition was underway in international politics (from the US and West towards 'Asia').[62] Nevertheless, the idea that US unipolarity was central to this nuclear world remained strong. As late as 2019, Christopher Fettweis suggested that the problems at the heart of the Second Nuclear Age were driven as much by the 'absence of a superpower to balance the United States' and that this 'basic structural dynamic would lead to a number of unpleasant outcomes'.[63]

Scholars would continue to use the Second Nuclear Age framing to contextualise their analyses of global nuclear challenges for most of that decade. In 2014, prominent scholar and later US assistant secretary of defence, Vipin Narang warned that,

> The world has entered a second nuclear age. This age is not defined by a bipolar global superpower competition involving massive nuclear arsenals with the capability to destroy each other multiple times over. In this new era, regional nuclear powers will define the proliferation and conflict landscape. These states have small nuclear arsenals, are often ensnared in long-standing rivalries, participate in multiple conflicts, and often have weak domestic institutions.[64]

The Second Nuclear Age concept has continued into the Third Nuclear Age—indeed, for some the current period is better thought of as an extension of the Second Nuclear Age—with its existing attendant challenges—rather than something new. In 2018, Evan Braden Montgomery wrote of 'Sources of Instability in the Second Nuclear Age'.[65] Albeit the edited volume in which this chapter appeared engaged with a number of themes that

traded away for a world in which a brief race to reacquire nuclear weapons could become every former nuclear states overriding preoccupation.' Thomas Schelling, 'A world without nuclear weapons', *Daedalus* 138:4 (2009), p. 129.

[62] This was a prominent argument in Bracken, 'The second nuclear age', pp. 146–56.

[63] Christopher J. Fettweiss, 'Pessimism and nostalgia in the second nuclear age', *Strategic Studies Quarterly* 13:1 (Spring 2019), pp. 12–41.

[64] Vipin Narang, *Nuclear strategy in the modern era* (Princeton, NJ: Princeton University Press, 2014), p. 1.

[65] Evan Braden Montgomery, 'Sources of instability in the second nuclear age', in Lawrence Rubin & Adam N. Stulberg (eds.), *The end of strategic stability: Nuclear weapons and the challenges of regional rivalries* (Washington, DC: Georgetown University Press, 2018).

I argue represent the move into a Third Nuclear Age, and reflected significance discontinuities with earlier framings.[66] In 2024, Paul Bracken wrote of 'Navies in the Second Nuclear Age',[67] and in 2025 Jan Ludvic at the Peace Research Centre in Prague discussed whether we might be in the 'second and a half nuclear age'.[68] A year later, the highly respected Italian historians Leopoldo Nuti and Giordana Pulcini argued in 'Un Deuxième—et un Troisième—Âge Nucléaire?'[69] that the current period of nuclear history is better understood as a long transition that can be traced back to the 1970s. I address this in the next section.

This 'Second Nuclear Age' literature increasingly included scholars not based in the West, and writing about the non-Western world—suggesting that what had begun as a Western concept had slowly become normalised.[70] Or at least it reflected the extent to which scholars saw this particular framing as a necessary way into what was—and arguably still is—a debate dominated by a handful of Western voices, periodicals, and outlets. Either way, while there is no convenient demarcation point as with the shift from a First to Second Nuclear Age, there is a general consensus that the main challenges facing the global nuclear order have shifted again, and that we are already in a Third Nuclear Age.

The key to understanding this is in the dominance of Western and primarily US-based thinking, policy, and scholarship on the global debate. In effect, a small group of scholars, commentators, and policy officials brought about this change through the way that they spoke about the nuclear world. For sure, this period saw material changes in international politics, but the shape and scope of the post-Cold War nuclear context and how it could and should be understood was in some ways up for grabs in the 1990s. Ultimately, this battle of ideas was 'won' by public and political figures in the US who shaped a particular view of what global nuclear order was and should be.[71] The net result is that the nuclear story so far largely reflects Western

[66] Lawrence Rubin & Adam N. Stulberg, 'Introduction', in Rubin and Stulberg, *The end of strategic stability*, pp. 4–8.

[67] Paul Bracken, 'Navies in the second nuclear age', *Orbis* 68:1 (2024), pp. 58–71.

[68] Jan Ludvik, 'Third or "second and a half"? Continuity, path dependence, and the Third Nuclear Age', *Global Policy* 16 (2025), pp. 934–944.

[69] Leopoldo Nuti & Giordana Pulcini, 'Un deuxième—et un troisième—âge nucléaire? Réflexions sur la périodisation de l'histoire nucléaire', *Relations Internationales* 201:1 (2025), pp. 237–52.

[70] E.g. Toshi Yoshihara & James R. Holmes (eds.), *Strategy in the second nuclear age: Power, ambition and the ultimate weapon* (Washington, DC: Georgetown University Press, 2012).

[71] A theme addressed in, Campbell Craig, Jan Ruzicka, & Kjølv Egeland, 'A subtle call to greatness: Unipolarity, US foreign policy, and the indefinite extension of the NPT', *European Journal of International Security* 10:2 (2025), pp. 312–29.

constructs and beliefs, not necessarily global ones, largely because of the reality of the material and intellectual power and dominance of the US.

As is discussed below, there is more than enough evidence of fundamental discontinuity in our nuclear world today to suggest that a new era is upon us. But while the First Nuclear Age died abruptly, perhaps the past decade is better thought of as the slow death of the Second Nuclear Age.[72]

Atomic semantics: What is a Third Nuclear Age?

The notion that the world has moved into a 'New' or 'Third' Nuclear Age, and that the nuclear challenges facing us today are different from the past, has rapidly gained traction in international academic, non-governmental organisation (NGO), media, and policy circles. However, the majority of the intellectual contributions to theorising this shift remain bounded within the limited, largely Western-centric viewpoint of the canon of First and Second Nuclear Ages literatures. Specifically, the challenges posed to (US) deterrence and strategic stability as a result of great power nuclear competition dominate the analysis. These challenges are important—particularly how the literature analyses ongoing shifts in the US national security zeitgeist given the impact that this continues to have on all of us on the planet. However, it fails to see the broader picture, and to seek to understand the Third Nuclear Age as a global phenomenon that goes beyond the US-Russia-China great power nuclear deterrence narrative, the nine states armed with nuclear weapons and the handful of regimes that might be seeking this capability.[73] Consequently, a modified and expanded conceptual framework is needed for the notion of a Third Nuclear Age to have traction and impact; specifically one that looks more broadly at the nuclear condition, how it is understood, and assesses what is changing, how much, in what ways, for whom, and with what implications.

Speculation about a possible Third Nuclear Age began as far back as 1999. As Colin Gray put it at the time:

> On current form in world politics, though not yet quite predictably, a putative third nuclear age is most likely to be organised by the return of a

[72] This phrase is still on a whiteboard in the room used by the Third Nuclear Age team at the University of Leicester as we struggled to conceptualise and represent what we all agreed was a changing nuclear world ...

[73] This is something addressed at depth in Andrew Futter, Ludovica Castelli, Cameron Hunter, Olamide Samuel, & Francesca Silvestri, *The global Third Nuclear Age: Clashing visions for a new era in international politics* (London: Routledge, 2025).

> single central axis of antagonism in world politics ... By far the most probable future central axis of antagonism in world politics is one between the United States and China.[74]

But the modern usage of the term began to appear around the mid- to late 2010s, alongside a more broadly used concept of a 'new nuclear age'. Both expressions recognised that the nuclear world had begun to look very different to the period that immediately followed the end of the Cold War, as a mixture of different and sometimes competing pressures were seen as transforming the geopolitical landscape, the types of threats that we worried about, and the scope of the debate. Both remained largely undertheorised and narrowly conceived; but how we think about the Third Nuclear Age matters.

In 2010, Thomas Donnelly and David Trachtenberg warned that 'The arrival of the third atomic age is no happy dawn'.[75] In 2018 Czech scholar Michel Smetana wrote of 'A Nuclear Posture Review for the Third Nuclear Age',[76] and in 2019 Jenny Naylor at Missouri State University argued that 'technology-enabled, multipolar competition will usher in the third nuclear age'.[77] A few years later, former US Office of the Secretary of Defense official David Cooper wrote a book about *Arms Control for the Third Nuclear Age*;[78] in 2023 Indian analyst Joy Mitra published his monograph *Conventional Military Strategy in the Third Nuclear Age*,[79] and French expert Emmanuelle Maitre authored a report titled, 'La Dissuasion au Troisième Âge Nucléaire'.[80] In 2024, German researchers Lucian Bumeder and Tobias Fella published an article looking at 'Germany in the Third Nuclear Age'.[81]

In a quite different historical conceptualisation, the Carnegie Endowment's Ariel Levite pointed to the period of 'complacency and disillusionment' between 1993 and 2010 as the Third Nuclear Age, and suggested that what would follow would be a Fourth Nuclear Age. Many components of

[74] Gray, *The second nuclear* age, p. 7.

[75] Donnelly & Trachtenberg, 'Toward a new "new look"', p. 1.

[76] Michel Smetana, 'A nuclear posture review for the Third Nuclear Age', *The Washington Quarterly* 41:3 (2018), pp. 137–57.

[77] Jenny Naylor, 'The third nuclear age', *Comparative Strategy* 38:4 (2019), pp. 276–88.

[78] David Cooper, *Arms control for the Third Nuclear Age* (Washington, DC: Georgetown University Press, 2022).

[79] Joy Mitra, *Conventional military strategy in the Third Nuclear Age* (London: Routledge, 2023).

[80] Emmanuelle Maitre, 'La dissuasion française au troisième âge nucléaire', Fondation pour la Recherche Stratégique, 10 October 2203, https://www.frstrategie.org/publications/recherches-et-documents/dissuasion-francaise-troisieme-age-nucleaire-2023.

[81] Lucian Bumeder & Tobias Fella, 'Germany in the Third Nuclear Age', *Internationale Politik Quarterly*, 11 September 2024, https://ip-quarterly.com/en/germany-third-nuclear-age.

Levite's Fourth Nuclear Age are similar to the more dominant framing of a Third/New Nuclear Age: new proliferation challenges, geopolitical competition, and possible confrontation, as well as a reinvigorated push for disarmament.[82]

By the early 2020s, the idea of a Third Nuclear Age had begun to get traction in professional and policymaking circles too. In October 2024, *The Economist* wrote of the 'perils of the world's third nuclear age',[83] and Indian member of parliament (MP) Manish Tewari wrote about 'India's Role in the Third Nuclear Age' in *The Indian Express.*[84] In 2024, North Atlantic Treaty Organization (NATO) supreme allied commander for transformation, Admiral Pierre Vandier's *La Dissuasion au Troisième Âge Nucléaire* (published in French in 2018) was translated into English as *Deterrence in the Third Nuclear Age.*[85] In 2022, director of the Defense Policy Laboratory of the Japanese Ministry of Defense Research Institute, Sugio Takahashi, wrote about 'The Beginning of a Third Nuclear Age'.[86] And in 2025, the Norwegian Nobel Institute hosted a symposium focussing on 'The challenges of the Third Nuclear Age'.[87]

However, with a few exceptions, such as the article that I wrote with Ben Zala in 2021,[88] the *Global Third Nuclear Age* book that my research team and I published in 2025,[89] and the University of Glasgow's Rhys Crilley in *Unparalleled Catastrophe* in 2023,[90] the term Third Nuclear Age continued to be used unproblematically and rarely explained, and contained a very Western, if not US, focus/bias.[91]

[82] Levite, 'Heading for the fourth nuclear age', p. 25.

[83] Anton La Guardia, 'The perils of the world's third nuclear age', *The Economist*, 20 November 2024, https://www.economist.com/the-world-ahead/2024/11/20/the-perils-of-the-worlds-third-nuclear-age.

[84] Manish Tewari, 'India's role in the third nuclear age', *The New Indian Express*, 31 October 2024, https://www.newindianexpress.com/opinions/2024/Oct/31/indias-role-in-the-third-nuclear-age.

[85] Pierre Vandier, *La dissuasion au troisième âge nucléaire* (Gastaldi: Editions du Rocher, 2018); translated by Armel Dirou & Julien Lalanne de Saint Quentin, *Deterrence in the Third Nuclear Age* (London: Palgrave MacMillan, 2025).

[86] Sugio Takahashi, 'The beginning of the Third Nuclear Age', *Yomiuri Shimbun*, 24 March 2022, https://www.yomiuri.co.jp/commentary/20220323-OYT8T50176/.

[87] 'Managing the challenges of the Third Nuclear Age', Symposium, Norwegian Nobel Institute, 2025. The product of this symposium will be an edited book titled, *Managing the challenges of the Third Nuclear Age.*

[88] Andrew Futter & Benjamin Zala, 'Strategic non-nuclear weapons and the onset of a third nuclear age', *European Journal of International Security* 6:3 (2021), pp. 257–77.

[89] Futter et al., *The global Third Nuclear Age.*

[90] Rhys Crilley, *Unparalleled catastrophe: Life and death in the Third Nuclear Age* (Manchester: Manchester University Press, 2023).

[91] Indeed, some have argued that the label was/is being used instrumentally. See Tom Vaughan, 'The "third nuclear age" is a politically motivated label that seeks to justify a renewed

In a broadly similar time frame, others have written about the 'new' nuclear age—mostly but not always linking this with the ostensible shift in US strategic thinking away from 'rogue states' towards the challenges posed by China and Russia. While this might seem like semantics: the labelling of New versus Third is important and instructive. Perhaps most pointedly, *New* suggests a rupture point and perhaps overwhelming discontinuity, whereas a *Third* Nuclear Age suggests a more gradual and multifaceted transition. It is also interesting to note that the New Nuclear Age literature is almost exclusively focussed on the US and written by scholars and professionals in the US. It is also largely focussed on US deterrence, great power competition, and the challenges posed by Russia and China. Good examples of this are the collection of essays edited by Robert Legvold and Christopher Chyba in *Daedalus* in 2020,[92] Linton Brooks, Francis Gavin, and Alexei Arbataov, *Meeting the Challenges of the New Nuclear Age* in 2018;[93] Steven Miller, Robert Legvold, and Lawrence Freedman, *Nuclear Weapons in a Changing Global Order*;[94] Scott Sagan and Vipin Narang's *Fragile Balance of Terror* in 2023,[95] Ankit Panda's, *The New Nuclear Age*,[96] and an article by Vipin Narang and Pranay Vaddi unpacking how (the US) could survive the new nuclear age, both published in 2025.[97] One notable exception to this American dominance is Ichimasa Sukeuki's edited collection published by the Japanese Institute for Defense Studies, *New Horizons of the Nuclear Age* in 2024.[98]

arms race', *The Conversation*, 15 August 2025, https://theconversation.com/the-third-nuclear-age-is-a-politically-motivated-label-that-seeks-to-justify-a-renewed-arms-race-263009.

[92] Robert Legvold & Christopher Chyba, 'Meeting the challenges of a new nuclear age', *Daedalus* 149:2 (2020), pp. 6–237.

[93] Linton Brooks, Francis J. Gavin, & Alexei Arbatov, *Meeting the challenges of the new nuclear age* (Cambridge, MA: American Academy of Arts & Sciences, 2018).

[94] Robert Legvold & Lawrence D. Freedman, *Nuclear weapons in a changing global order* (Cambridge, MA: American Academy of Arts & Sciences, 2018).

[95] Scott Sagan & Vipin Narang (eds.), *The fragile balance of terror: Deterrence in the new nuclear age* (Ithaca, NY: Cornell University Press, 2023).

[96] Ankit Panda, *The new nuclear age* (Cambridge: Polity Press, 2025). As Panda describes it: 'The decade beginning in 2020 marks a transition to a new, more dangerous, and more complex third nuclear age: a period that will be defined by the renewal of great power nuclear competitive dynamics, complex multipolar nuclear deterrence relationships, newly arrived and increasingly sophisticated nuclear actors like North Korea, and the anxieties introduced by the still-uncertain effects of a range of emerging technologies (such as hypersonic weapons, cyberweapons / artificial intelligence, and possible space weapons). The challenges of this new nuclear age will intensify in ways that that are unlikely to be fully foreseeable as so-called "great power competition" between the U.S., Russia, and China intensifies.'

[97] Vipin Narang & Pranay Vaddi, 'How to survive the new nuclear age: National security in a world of proliferating risks and eroding constraints', *Foreign Affairs*, July/August 2025, https://www.foreignaffairs.com/united-states/how-survive-new-nuclear-age-narang-vaddi.

[98] Ichimasa Sukeuki (ed.), *New horizons of the nuclear age* (Tokyo: National Institute for Defense Studies, 2024).

The idea of a 'new' nuclear age suggests a major rupture point and possibly revolutionary change. But across this literature there is little consensus about when the Third Nuclear Age may have started. Rhys Crilley points to the election of Donald Trump in 2016 and the US decision to leave the Intermediate-Range Nuclear Forces (INF) Treaty in 2019,[99] and Ankit Panda sees the materialisation of the US-China nuclear rivalry with the discovery of over 120 missile silos in 2021 as a key inflection point.[100] In a similar vein, Nicholas Miller and Vipin Narang noted that, 'future scholars may look back on 2019 as the turning point from an era of relative calm to one of intense nuclear competition and proliferation—the dawn of a dangerous new nuclear age'.[101] But as is explained below, searching for a specific moment may obfuscate a gradual and somewhat staccato change that has probably been underway for much longer.

While the literature on a new or Third Nuclear Age is sporadic and of course in its infancy, two significant themes have emerged. The first is a certain constriction in terms of focus—the vast majority of work focusses on a combination of nuclear competition and stability between the US, Russia, and China, the impact of disruptive technologies on deterrence, and an end to arms control and restraint. The second is that these dynamics emerged in the mid-2010s and represented a major shift—possibly even a transformation—in a certain version of international politics. Both of these are clearly important components for understanding today's nuclear world, but provide a rather constrained view of what the Third Nuclear Age is and why. The argument made here is that to properly understand the Third Nuclear Age one must do two things: appreciate that change in global nuclear politics is complicated, experienced differently, and often messy, and that nuclear order has multiple different constituent components and stakeholders beyond the great powers.

How we seek to understand change is central to conceptualising global nuclear politics. Change across our nuclear world can happen in different ways, at different speeds, and be experienced differently. With this in mind, perhaps the first thing to note is that the shift into a Third Nuclear Age is not absolute—clearly important dynamics endure from previous eras. As Robert Levgold and Christopher Chyba explain,

[99] Crilley, *Unparalleled catastrophe*, p. 1.
[100] Panda, *The new nuclear age.*
[101] Nicholas Miller & Vipin Narang 'Is a new nuclear age upon us?', *Foreign Affairs*, 30 December 2019, https://www.foreignaffairs.com/united-states/new-nuclear-age-upon-us.

> the dangers of the earlier nuclear era remain: some familiar, some in altered form, and to them are added new ones. The risk of inadvertent nuclear war remains, but now the pathways to it have multiplied across more regions and relationships ... The Cold War potential for misreading the other side's nuclear thinking and plans swells when it extends to a larger and diverse set of nuclear actors. The accumulated conceptual refinement and residual understanding on nuclear issues built between the United States and the Soviet Union were the result of fifty years of strategic nuclear arms negotiations and dialogue; this shared understanding has no modern counterpart in any other bilateral or trilateral nuclear relationship.[102]

The challenges of the First Nuclear Age didn't completely disappear in 1991 and neither did the Second Nuclear Age concerns about nuclear terrorism and rogue actors at some unspecified date in the early 2010s. The threat that non-state actors could gain access to nuclear material, that North Korea may use nuclear weapons, or that Iran (or others) might acquire them is still significant. The decision for the US to bomb Iranian nuclear sites in June 2025 (and to escalate attacks 8 months later) is a good example of this. At the same time, what might be thought of as First Nuclear Age challenges such as the possibility of horizontal nuclear proliferation to new actors, the threat of a conventional interstate war that leads to major nuclear exchange, and the fact that nuclear war can still destroy our modern world and perhaps also the planet, have returned to prominence too.

The cut-off point of the end of the Cold War might suggest a rather abrupt change; in reality the First and Second Nuclear Ages, and, indeed, the Third, slowly 'bled' into one another.[103] Four examples are worth noting here. First, many of the proliferation challenges central to the Second Nuclear Age began in the First Nuclear Age (albeit not explicitly for what might be thought of as First Nuclear Age reasons—essentially the Cold War standoff): India and Pakistan may well have had undeclared nuclear weapons capabilities by the end of the 1980s, and Iraq, North Korea, Libya, and others had nuclear programmes underway before the Berlin Wall came down.[104] Second, the decisions taken about how best to respond to the challenges of

[102] Robert Levgold & Christopher Chyba, 'Introduction: The search for strategic stability in a new era', *Daedalus* 149:2 (Spring 2020), p. 9.

[103] This was a notion that Ben Zala and I first explored in Futter & Zala, 'Strategic non-nuclear weapons'.

[104] See Mahdi Obeidi & Kurt Pitzer, *The bomb in my garden: The secrets of Saddam's nuclear mastermind* (Nashville, TN: Turner Publishing Company, 2008); Målfrid Braut-Hegghammer,

the Second Nuclear Age—namely how to address 'rogue' and 'undeterrable' nuclear actors through precision strike weapons and missile defences, and later US unilateralism and pre-emption—drove the development of doctrine and technological deployments that would fundamentally shape the contours of the Third Nuclear Age. Third, the disarmament activism that would manifest in the 2017 Treaty on the Prohibition of Nuclear Weapons can probably be traced to the early 1990s (if not much earlier) as scholars began theorising 'human security' and the gradual realisation that the decision to indefinitely extend the NPT would not fundamentally change the role of nuclear weapons in security politics. Fourth, the vision of a globalised, liberal, and democratic world order pursued by the US and its allies in the Second Nuclear Age sowed the seeds of discord that would result in the direct challenges to US primacy in the Third Nuclear Age. As Colin Gray put it back in 1999, 'There is no certainty about the general character, let alone detail, of the age after the current one, but much of the material from which that future age will be fashioned is around us now'.[105]

But the difference is that the nuclear challenges that animated the 1990s and 2000s no longer grab headlines and drive policy debates like they did in the past (at least in the West—such Second Nuclear Age concerns are still prominent in other parts of the world). This blending of nuclear ages is a reason why there is a less obvious cut-off point between the Second and Third Nuclear Ages, or at least not as clear as the end of the Cold War. What effectively happens is that hierarchies of threat—shift.

Moreover, understanding change in nuclear politics also means recognising that things happen at different speeds—sometimes this reflects something of a punctuated equilibrium,[106] but rarely are there distinct moments of 'rupture'; that the rate of change varies across different factors/dynamics; and what constitutes change and its implications is different for different actors. It is also important to appreciate that change is sometimes driven or shaped by the most powerful, either deliberately or unintentionally.

Indeed, First and Second Nuclear Age framings were shaped by powerful actors and reflected the interests of those actors. But to really understand the new period in our nuclear history it is essential to broaden our intellectual

Unclear physics: Why Iraq and Libya failed to build nuclear weapons (Ithaca, NY: Cornell University Press, 2016); and Jonathan D. Pollack, *No exit: North Korea, nuclear weapons, and international security* (London: Routledge, 2017).

[105] Gray, *The second nuclear age*, p. 7.

[106] A concept borrowed from biology which suggests long periods of relative stasis followed by short bursts of change.

horizons to look beyond this forgivably limited focus. Of course, broadening anything too far to include myriad dynamics and actors risks making it meaningless or tautological, but ignoring too many dynamics, factors, and voices in pursuit of a single parsimonious theory or model is arguably far more problematic.[107] A good reason for this is that global nuclear politics is not experienced by all stakeholders and actors in the same way. Appreciating that how we understand nuclear history and construct meaning of the current era and environment in which we live has implications and is often—instrumentally or not—the result of dominant socio-political forces. This means looking at what narratives of nuclear threats and nuclear order have been and are prioritised as well as ignored or downplayed and recognising the realities of power. There are four components to this.

First, it is important to look beyond the US-Russia-China great power competition framing, those states that currently have nuclear weapons or might possibly seek to acquire them, to include all actors within the global nuclear order. This means not just nation states, and the interests of 'middle powers' and the majority from the non-nuclear-armed 'global south', but across other levels of analysis that includes activists and NGOs, individuals, and those that construct nuclear knowledge and what constitutes nuclear reality. This also allows for an appreciation of the different ways in which different actors set about achieving nuclear security goals, how they are impacted by nuclear politics, and the incorporation or utilisation of nuclear technologies more broadly. While power and the macro-structure of the nuclear world are important, the nuclear story is also inherently a very human story.

Second, technological challenges are not solely about nuclear weapons and nuclear proliferation—although both are clearly central to global nuclear politics. What is often missing is an appreciation of nuclear technologies more broadly, and the various different non-weapons applications, particularly energy generation but also scientific research. Equally, we cannot understand nuclear weapons and changes and challenges to deterrence, arms control, and stability without placing these alongside developments in conventional weapons, and particularly what might be termed strategic non-nuclear weapons, and the wider digitised nuclear ecosystem.

Third, nuclear threats and nuclear technologies are part of a wider interconnected existential risk society. It is increasingly difficult to separate out

[107] As David Blagden reminded me, the late great Kenneth Waltz warned that a theory of everything is a theory of nothing; a perfectly detailed map is useless for navigation because it merely replicates the world.

nuclear risks from the major challenges posed to the human race by climate change and sustainable economic development, unprecedented interconnectedness, and the broader issues of global fairness. Nuclear technologies—especially if used for nuclear war—can impact the climate; a changing climate can impact nuclear security, safety, and perhaps proliferation; nuclear energy may be key to carbon-free development; and access to energy is a key component of global fairness. It is therefore important to think of the nuclear challenge as having planetary-level implications and as a component of wider deliberation about the Anthropocene.

Fourth, moving beyond (but not ignoring) the overly Western and US-centric focus of the canon of nuclear literature and putting this in context, instrumentally or otherwise, much of what we think we know about, and what we focus on, in nuclear politics is shaped by the considerable intellectual power and dominance of Western ideas and English-language scholarship. But the nuclear world looks different depending on where you find yourself on the geopolitical map, stepping back from focussing on quite specific debates about deterrence and disarmament to consider the broader intellectual framework within which this is happening. How we teach, learn about, and engage with nuclear politics will likely have considerable impact on the climate of ideas and what is possible in terms of our nuclear human future, in the medium to long term.

The final key component of this Third Nuclear Age framework focusses on causation. There is no one obvious driver of change into the Third Nuclear Age—and this is probably why it is so difficult to pinpoint a specific moment of transition. This is also why mono-causal models of capturing what matters in the nuclear world at any one time are limited in their explanatory power. It is essential to look at how different pressures and trends interlink, co-constitute, and clash. The relationship between factors that shape global nuclear politics is multi-causal, complex, and not linear.

When the nuclear world is looked at through this lens, several significant dynamics stand out that perhaps mark the transitional dynamic and departure from the Second Nuclear Age.

Technological change was important to the Second Nuclear Age, but this was primarily about decisions made about platforms, mission sets, and the requirements of deterrence by elites in the US. In the Third Nuclear Age, many actors are pursuing a range of different technologies that could have strategic effects, many of them non-nuclear and non-kinetic. This, along with a broader change in the techno-political environment within which we all live, creates a far more complex nuclear context within which operations and geopolitics play out.

The world today is increasingly competitive, with clashing visions of how international politics should be structured. This is shadowed by the possibility of moving towards *nuclear* multipolarity, characterised by a gradual diffusion and dispersal of power. While the US remains the most powerful material actor in the nuclear system, this is a long way from US hyperpower at the beginning of the 21st century. At the same time, it is not the same as the broadly bipolar structure of the Cold War. This shift in the global power balance has been combined with a growing appetite for nuclear risk acceptance and a growth of aggressive nuclear rhetoric, notably in and around the Ukraine War, but also linked to simmering conflicts in the Middle East, and northeast and south Asia. The nuclear taboo and norm of non-use were central components of the Second Nuclear Age but these now both seem to be weakening at the same time as we are seeing the enhanced likelihood of conventional interstate war.

The Third Nuclear Age will be far more global in scope than the Second Nuclear Age, and manifest as a more shared experience. The Second Nuclear Age construct was overwhelmingly centred on the US and a handful of rogue challengers to the US-led order. The Third Nuclear Age will be shaped by competition between the major powers and by the actions of middle powers, and nuclear politics in different parts of the world will be increasingly interlinked and perhaps magnified. This does not mean that current power structures will necessary be overturned, but that they will likely come under more scrutiny and pressure.

After a sustained period of global nuclear weapons reductions, pressures for current nuclear-armed states to build up their forces and deploy more nuclear weapons have returned. At the same time, the horizontal proliferation challenge has moved beyond 'rogue' challenges to include a number of middle powers, especially those reliant on US extended deterrence guarantees. In some ways this mirrors the early First Nuclear Age, only this time it is less clear who will champion non-proliferation and restraint.

Over the past two and half decades we have seen the gradual decaying and fragmenting of global nuclear institutions and frameworks, mainly built in the First Nuclear Age but that would become central to the Second Nuclear Age. The NPT and nuclear reductions though arms control were a key feature of the Second Nuclear Age construct, but have now begun to weaken. The agreement of the Treaty on the Prohibition of Nuclear Weapons (TPNW) is significant in this regard because it has sought to chart a nuclear future outside—or perhaps better thought of as beyond—the one previously devised by the nuclear-armed states, driving right at the heart of dominant conceptualisations of stability and nuclear order. Even more than this, the

TPNW is reflective of a much more conspicuous divide between deterrers and disarmers and the most desirable pathway to nuclear security for all members of the global community. This has replaced a certain harmony and optimism that had emerged in the early 1990s.

Finally, the 1990s and 2000s probably marked a nadir for interest in and demand for nuclear power, but a renewed push for nuclear technologies for various purposes is underway in the Third Nuclear Age. This includes both countries that previously chose to reduce or retire nuclear power and many developing states looking to join the nuclear power club. It also includes a role for powerful multinational corporations. This does not necessarily equal more weapons proliferation but will almost certainly create a number of security and safety challenges.

Conclusion: Continuity and change in global nuclear politics

The title of this chapter reflects an awkward attempt to make a pun from combining the political scientist Alexander Wendt's influential theory that 'anarchy is what states make of it'[108] with the long-running US TV show 'Whose Line Is It Anyway?'. But in a way both get to a pretty fundamental point about how we engage with and understand the nuclear condition: how we think about nuclear politics is constructed, with the most compelling and convincing version effectively winning the points.

This is because decisions about how we choose to periodise and conceptualise history, including nuclear history, are inherently political and subjective, and often reflective of dominant power structures—although this doesn't mean they are wrong or unhelpful. For better or worse the canon of nuclear history and the way that the evolution of nuclear order has been understood and chronicled—with a few notable exceptions—has been an inherently Western phenomenon, with a strong centre of gravity in the US. This reflects a certain English language and Atlanticist cultural dominance—and the reality that the US has been the foremost force in international politics and the security studies academy since 1945. The natural result is that the standard account of nuclear history is that the periods before and after 1990 were distinctly different.

[108] Alexander Wendt, 'Anarchy is what states make of it: The social construction of power politics', *International Organization* 46:2 (1992), pp. 391–425.

The moment that we started to move into a Third Nuclear Age is of course far from clear cut. The transition arguably began gradually during the 2010s, with no one obvious shared moment of inflection. But in many ways the transition is far longer, given the lingering dynamics that continue from both the First and Second Nuclear Age, today. Moreover, different dynamics have moved at different speeds and with different implications for different actors, which means that this experience isn't necessarily homogenous. But what is perhaps different to the previous two historical framings is that this transition has been and will continue to be a much more globally collective experience. While language may differ, and the implications of change may be differentiated, there is a feeling that the contours and nature of the global nuclear order are shifting—and that this is the result of technological, geopolitical, and normative change—or, perhaps more correctly, greater discontinuity than continuity across nuclear politics. There is also a feeling that most of this change—though not necessarily all—is negative or at least challenges established and countervailing (but also contested) notions of a stable and peaceful nuclear order.

But if we genuinely want to understand the current transformation and friction across and within the global nuclear order, and at the same time make the concept of a Third Nuclear Age internationally and intellectually viable, then the frame of reference must be expanded. Great power competition, deterrence, stability, and arms control remain essential components of this order, but they are just the pre-eminent elements of a multifaceted picture involving geopolitical change, the diverse ambitions of different stakeholders, and the growing normative and intellectual challenges to nuclear orthodoxy. Above all, we need to escape the implicit (and largely unconscious) US/Western ethnocentric lens that has been used to frame nuclear history so far.

The Third Nuclear Age construct is therefore an imperfect but also, I hope, fundamentally useful vehicle with which to seek to understand the nature and challenges of the changing nuclear world around us. It is to a central component of this nuclear story—technology—that we turn next.

2

Swords, shields, and the security dilemma

Introduction: Technology as a driver of change

The British science fiction pioneer Arthur C. Clarke once famously warned that 'Any sufficiently advanced technology is indistinguishable from magic'.[1] But at the heart of the Third Nuclear Age thesis is a belief that rapid technological change—often combined under the amorphous 'emerging and disruptive technology' (or EDT) label—is impacting nuclear politics in a range of different, but mostly negative, ways. The EDT moniker is habitually undertheorised and regularly used by different people to mean quite different things, and this is one reason why technological change is often hyped and misunderstood in the nuclear space.[2] But technological change is important to the Third Nuclear Age—not least the diffusion of different types of capabilities that might have strategic effects, to more actors. Indeed, in early 2021 Ben Zala and I published an article in the *European Journal of International Security* arguing that technological change—especially the rise of 'strategic non-nuclear weaponry'—would be a fundamental driver of the move into a Third Nuclear Age. As we wrote:

> We stand on the cusp of a Third Nuclear Age. If this new era comes to pass, the main characteristics of the global nuclear order are likely to see considerable challenge. This shift is being driven by the development of a suite of strategic non-nuclear weaponry (SNNW) and enabling technologies, in

[1] One of Clarke's 'three laws': Arthur C. Clarke, *Profiles of the future: An inquiry into the limits of the possible* (London: Harper & Row, 1973).

[2] Andrew Futter, 'Disruptive technologies and nuclear risks: What's new and what matters', *Survival* 64:1 (2022), pp. 99–120. Also thoughtfully addressed by Marina Favaro, in 'Weapons of mass distortion', Kings College London, 2021, https://www.kcl.ac.uk/csss/assets/weapons-of-mass-distortion.pdf.

The Third Nuclear Age. Andrew Futter, Oxford University Press. © Andrew Futter (2026).
DOI: 10.1093/9780198944126.003.0003

> conjunction with a fundamental political shift relating to perceptions of nuclear threats and debates about how they should be addressed.[3]

Half a decade later,[4] I think we were right to make the case for the centrality of new types of weapons and support systems being at the heart of a transformation in our nuclear world—after all, from the development of the printing press in the 1400s, to the Industrial Revolution in the mid-18th century, and the first nuclear bomb in 1945, technology has always driven societal, political, and military change. But what perhaps we didn't get quite right was the messy, uneven, and often obfuscated nature of this impact and transformation.

Part of the reason for this is because what we might think of as technological change in the Third Nuclear Age is multifaceted. First, technological change is not just about weapons and support systems but also about the information context within which nuclear operations and nuclear politics play out. Second, it involves a mixture of weapons platforms and support systems that have evolved, some capabilities that may be able to carry out specific mission sets that are different from before, some technological applications that have effectively been rediscovered and aren't really that new at all, and some technological developments that create the possibility of having strategic effects on nuclear operations in ways that are different from the past. Third, it also involves—as the next chapter explains—the development and deployment of particular capabilities for different political, military, and sometimes cultural purposes.

The net result is a more complex techno-political environment within which nuclear politics plays out, and potentially more fuel to the fire of security dilemma dynamics. Of course, fears of technological change have always been part of the nuclear story, especially worries that some technological breakthrough might somehow undermine deterrence—the most worrying being the creation of the possibility of a disarming first strike. But perceptions of technological change, and particularly a fear that an adversary may be seeking some type of advantage, have—and can—also drive insecurity even without intent. As one of the earliest exponents of the theory of the 'security dilemma', the American scholar John Herz put it three-quarters of a century ago:

[3] Andrew Futter & Benjamin Zala, 'Strategic non-nuclear weapons and the onset of a Third Nuclear Age', *European Journal of International Security* 6:3 (2021), p. 257.

[4] And arguably a bit longer given that the article was rejected several times by other periodicals first ...

> Wherever such anarchic society has existed and it has existed in most periods of known history on some level-there has arisen what may be called the 'security dilemma' of men, or groups, or their leaders. Groups or individuals living in such a constellation must be, and usually are, concerned about their security from being attacked, subjected, dominated, or annihilated by other groups and individuals. Striving to attain security from such attack, they are driven to acquire more and more power in order to escape the impact of the power of others. This, in turn, renders the others more insecure and compels them to prepare for the worst. Since none can ever feel entirely secure in such a world of competing units, power competition ensues, and the vicious circle of security and power accumulation is on.[5]

The chapter therefore maps out the technological landscape of the unfolding Third Nuclear Age, placing recent developments in historical context. It distinguishes between the gradual yet significant technological shifts shaping nuclear politics and strategy today, and those that, while often portrayed as revolutionary, may be thought of as more marginal, exotic, or even quixotic. It also explains why technological change is not homogenous, the impact isn't equal, and that there is no one 'magic bullet' to blame.

Déjà vu all over again: Technological change in perspective

It can be easy to assume that the technological challenges at the heart of the Third Nuclear Age, and the way space weapons, artificial intelligence (AI), hypersonic missiles, and other advanced applications of science to warfare appear poised to reshape deterrence, stability, arms control, proliferation, and crisis management are somehow unstoppable and completely new and novel. But as my colleagues Bleddyn Bowen and Cameron Hunter have never tired of reminding me over the past few years, most of the capabilities, platforms, and mission sets currently being hyped as transformative have been around in one form or another for decades.[6] This is not to say that technological change and innovation haven't been central to the nuclear

[5] John H. Herz, 'Idealist internationalism and the security dilemma', *World Politics* 2:2 (1950), pp. 157–80.

[6] Indeed, in an episode of the hugely successful 1980s TV satire *Yes, Prime Minister*, the scientific advisor to Prime Minister Jim Hacker urges him to scrap Trident and spend the money on 'ET'. 'Not extra-terrestrials, but "emergent technologies"'. Series 1, episode 1: 'The grand design'.

story, or that such dynamics aren't significant today, but that a certain 'pinch of salt' needs to be taken when analysing these trends.

There is perhaps no better example of technological change transforming global politics than the invention of the nuclear bomb itself. As is discussed in the previous chapter, the 'nuclear revolution', especially when combined with miniaturised warheads, long-range, accurate and hard-to-counter delivery platforms, robust command and control systems, and thermonuclear physics packages, arguably altered the way that society has thought about peace and security ever since. Despite a desire by different actors to mitigate the condition of vulnerability or seek strategic advantage, or as Keir Lieber and Daryl Press describe it in their 2020 book 'escape nuclear stalemate'[7]—the basics of nuclear technology have remained largely unchanged across the First, Second, and Third Nuclear Ages.

There have of course been moments of panic and fears that certain technological leaps have or might transform the strategic landscape in the past. The launch of the Sputnik satellite by the Soviet Union in October 1957 played into notions in the United States (US) of a 'missile gap' and a window of vulnerability to intercontinental-range attacks that could hit targets in minutes rather than hours;[8] the deployment of the first missiles armed with multiple independently targetable warheads ('MIRV') in 1970 raised concerns about the ability to carry out a surprise 'first strike' that might disable an adversary's nuclear capability; and the announcement of the Strategic Defence Initiative by US President Ronald Reagan in 1983 caused huge panic in the Kremlin (as well as in Europe and elsewhere[9]) that nuclear arsenals could be rendered obsolete by defensive shields. But material technological change—as opposed to the grandstanding announcements and publicity that has often accompanied it—has rarely happened overnight or transformed the strategic equation in the way that some might have feared. While technological change has unquestionably been significant, it has therefore been largely evolutionary and punctuated rather than revolutionary.

[7] Keir Leiber & Daryl Press, *The myth of the nuclear revolution: Power politics in the atomic age* (Ithaca, NY: Cornell University Press, 2020).

[8] See chapter 11, 'Sputnik and the Soviet threat' in Lawrence D. Freedman & Jeffrey Michaels, *The evolution of nuclear strategy* (London: Palgrave MacMillan, 2019), pp. 169–78.

[9] For a discussion of this, see Aaron Bateman, *Technology in space: Technology, politics and the rise and fall of the Strategic Defense Initiative* (Cambridge, MA: MIT Press, 2024); and Aaron Bateman, 'Science, technology, and the "Special Relationship": A re-examination of Britain's support for the Strategic Defense Initiative', *Science & Diplomacy* 8:1 (May 2019).

Perhaps this is why many of the technological challenges that will shape the Third Nuclear Age can be tracked back many years if not decades. Attempts to intercept or shoot down cruise and ballistic missiles can be traced back to the Nazi V-bomb attacks against London during World War Two;[10] in the 1950s, British scientist Alan Turing and others began thinking about what would become known as AI,[11] eminent Soviet physicist Andrey Sakarov devised the idea of a multimegaton nuclear torpedo able to cause a Tsunami,[12] and the US Air Force Project A119 proposed detonating a nuclear weapon on the Moon;[13] in the 1960s, the Soviet Union developed a Fractional Orbital Bombardment platform,[14] the first direct-ascent anti-satellite weapons were tested,[15] and Project Pluto looked into the viability of the US developing nuclear-powered cruise missiles;[16] Andrew Marshall and colleagues at the Office for Net Assessment began looking at what would become conventional precision strike,[17] and the Pentagon commenced a programme to develop a 'nuclear-pumped x-ray laser' for missile defence[18] in the 1970s. After watching the 1983 Hollywood film *WarGames* President Reagan launched an investigation into security of the 'most sensitive US computer systems' to what would later be termed 'cyber attacks',[19] and a year later the Harvard scholar Matthew Bunn gave a detailed public overview

[10] See Richard Dean Burns & Lester H. Brune, *The quest for missile defenses, 1944–2003* (Mineola, NY: Regina Books, 2003).

[11] This is where we get the concept of the 'Turing test' about whether machines might be able to 'think'. Alan M. Turing, 'Computing machinery and intelligence', *Mind* 59:236 (1950), pp. 433–60.

[12] 'T-15 nuclear torpedo', GlobalSecurity.org, https://www.globalsecurity.org/wmd/world/russia/t-15.htm.

[13] E.g. Robin Andrew's, 'This is the story of how America once thought about nuking the moon', Forbes, 6 August 2018, https://www.forbes.com/sites/robinandrews/2018/08/06/this-is-the-story-of-how-america-once-thought-of-nuking-the-moon/.

[14] See Miroslav Gyurosi, 'Soviet fractional orbital bombardment program', Technical Report APA-TR-2010-0101, https://www.ausairpower.net/APA-Sov-FOBS-Program.html (updated April 2012).

[15] Laura Greco, 'A history of anti-satellite programs', Union of Concerned Scientists, January 2012, https://www.ucs.org/sites/default/files/2019-09/a-history-of-ASAT-programs_lo-res.pdf.

[16] See, for example, Barton C. Hacker, 'Whoever heard of nuclear ramjets? Project Pluto, 1957–1964', *Icon* 1 (1995), pp. 85–98.

[17] In fact, as James Acton notes, 'As far back as 1975 a RAND study raised the possibility of developing highly accurate guidance systems that might allow a conventional warhead to be used on an ICBM'. James Acton, *Silver bullet: Asking the right questions about conventional prompt global strike* (Washington, DC: Carnegie Endowment for International Peace, 2013), p. 10.

[18] See, for example, William J. Broad, *Teller's war: The top secret story behind the Star Wars deception* (London: Simon & Schuster: 1992).

[19] Fred Kaplan, *Dark territory: The secret history of cyber war* (New York: Simon and Schuster Paperbacks, 2016), pp. 1–2.

of 'manoeuvrable' (or what we might now term 'hypersonic') warheads in 1984.[20] At the end of the First Nuclear Age renowned physicist Kosta Tsipis published a book focussing on the strategic impact of 'new technologies' including 'materials, sensors, microelectric circuits and computers'.[21] The very hardware that much of the most sophisticated weapons systems and support platforms utilise today.

As we moved into the Second Nuclear Age and the geopolitical context shifted from superpower competition to 'rogue' actors, Michael Russel Rip and James M. Hasik were writing of the 'precision revolution'[22] as a result of new positioning, navigating, and timing (PNT) technologies, and Dennis Gormley warned about the impact of the proliferation of cruise missiles beyond the advanced Western states.[23] In 1991, the purported success of the Patriot missile defence system in intercepting Iraqi scud missiles was used by some to justify the push for strategic missile defences.[24] A decade later in 2002, the George W. Bush administration abrogated the Anti-Ballistic Missile (ABM) Treaty, which for 30 years had placed limits on the deployment of US and Russian ballistic missile defences citing a 'new strategic context' and a technological imperative.[25] At the same time, the US 'global prompt / precision strike' programme was launched—seeking to provide 'long-range' conventional strike options, possibly against nuclear targets.[26]

What is interesting is how we can trace the evolution of these capabilities and technologies and see how they have changed over time. But perhaps importantly, this genealogy shows us that it was rarely ever one specific technology that mattered, but combinations of them, and fears of worst-case future scenarios. The nexus between the US decision to deploy ballistic missile defences, enhance the ability to monitor possible targets from space (or cyberspace), and develop non-nuclear precision weapons to meet the Second Nuclear Age challenges posed by 'rogue' states, for example, was—and

[20] Matthew Bunn, *Technology of ballistic missile re-entry vehicles*, in Kosta Tsipis & Penny Janeway (eds.), *Review of US military research and development, 1984* (Cambridge, MA: Massachusetts Institute of Technology: 1984).

[21] Kosta Tsipis, *New technologies, defense policy, and arms control* (London: Harper & Row, 1989), p. 7.

[22] Michael Russel Rip & James M. Hasik, *The precision revolution: GPS and the future of aerial warfare* (Annapolis, MD: Naval Institute Press, 2002).

[23] Dennis M. Gormley, *Missile contagion: Cruise missile proliferation and the threat to international security* (Annapolis, MD: Naval Institute Press, 2008).

[24] See Theodore Postol, 'Lessons of the Gulf War experience with Patriot', *International Security* 16:3 (1991), pp. 119–71.

[25] For a discussion of this, see Andrew Futter, *Ballistic missile defence and US national security policy: Normalisation and acceptance after the Cold War* (London: Routledge, 2013).

[26] See Acton, *Silver bullet*.

continues to be—seen as a direct threat to nuclear surety in Moscow and Beijing today.[27]

When seeking to analyse the impact of technological change on nuclear politics it is important to differentiate specific technologies or groups of technologies (which might develop at different speeds and have very different applications), from the mission sets that these technologies may be developed to carry out, and from how they are understood by others. For example, missile defence is not one technology, but many different systems designed for different purposes, comprising detection and tracking sensors in different domains, data processing facilities, and kill vehicles / interception and guidance methods. The same is true of kinetic or non-kinetic precision strike attacks, anti-submarine warfare, counterspace operations, anti-access area denial zones (A2AD), defence suppression (suppression of enemy air defence, or SEAD), and any broader notions of counterforce or deterrence by denial (discussed in the next chapter). Indeed, quite often the literature and debate about technology focusses too much on the front-end part—that is the 'shooter' or 'effector' rather than the myriad systems that allow for a particular mission to be carried out.

Much like the nuclear ages literature, the story of technological challenges to nuclear order is also a very Western / great power, even bilateral, story, and this reflects the political and power realities of First and Second Nuclear Ages (as discussed in Chapter 1). For most of the period since 1945 it has been the US, for a time joined by the Soviet Union, that has been at the forefront of technological change.[28] One of the big differences in the Third Nuclear Age is that this is becoming less pronounced—technological diffusion now involves more actors in different parts of the world. Some of whom, notably China, potentially seek advantage against the US. This also means that technological timelines are not the same for all actors.

What this suggests is that while technological change is important to the Third Nuclear Age, we should not fall into the trap of assuming or believing that everything is new or different. To unpack this further: the technological challenge can be thought of as comprising three key dynamics: First it is about the diffusion of capabilities that in the First and Second Nuclear Ages were the preserve of a handful of developed nations, and principally the US, to more actors across the globe. Second, it is about how innovation is

[27] See, for example, Andrew Futter & Benjamin Zala, 'Advanced US conventional weapons and nuclear disarmament: Why the Obama plan won't work'. *The Nonproliferation Review* 20:1 (2013), pp. 107–22.

[28] And, for a brief period in the 1940s and 1950s, included the United Kingdom (UK).

lowering the barriers for more actors to engage in nuclear techno-politics (especially in the digital space), at the same time as emphasising lower tech options such as 'drones' or interference in the information ecosystem. Third, it is about how the major powers are utilising and adapting to technological change to create enhanced and in some cases new weapons platforms that facilitate expanded or different mission sets. None of these have emerged out of the blue, but all seem set to play a much bigger role in nuclear politics than in the past. At the same time, it is important to remember that technologies in and of themselves don't shape politics. As Keir Leiber and Daryl Press explain:

> scholars often overstate the degree to which technology shapes politics. National leaders and military planners are rarely confident enough in their understandings of how weapons and prevailing technologies will interact during war to be moved to radically alter their basic security strategies, much less their core political objectives.[29]

However, perceptions matter. And what seems to make the challenges posed today distinctive from the past is that many different technologies, weapons, and support systems are emerging at a similar time, and are being pursued by more actors than in the past, and this is unfolding in an increasingly digitised nuclear environment or 'ecosystem'.

Buzzword bingo: Beware the quantum, AI, hypersonic, space laser weapon ...

According to the generative AI platform ChatGPT, Buzzword Bingo is 'a humorous game played in meetings, conferences, or presentations where participants keep track of overused corporate jargon, technical terms, or trendy buzzwords'. And, at times, reading or engaging with the topic of 'emerging and disruptive technologies', especially as they might pertain to nuclear weapons, can feel a bit surreal. Everything, from various nuclear systems and delivery platforms, to deterrence, strategic stability, and arms control, seems to be vulnerable in lots of different ways to rampant technological innovations without any obvious means to mitigate these risks. However, and while there are undoubtedly several relatively new or emerging

[29] Liber & Press, *The myth of the nuclear revolution*, p. 126.

technologies that *could* possibly have an impact on nuclear politics and operations in the Third Nuclear Age, these are rarely clear cut or well understood. Five that are worth looking at in detail are 'hypersonic' missiles, 'cyber threats', AI, the weaponisation of space, and Quantum.

***

In some ways, the 'hypersonic missile' has become something of an emblem of the technological challenges of the Third Nuclear Age. Visions of extremely fast futuristic-shaped weapons, travelling at trajectories that make observation and tracking from the ground or the surface of the sea difficult, manoeuvring to defeat missile defences, and striking targets with high accuracy often without warning, seem to be at the heart of how technological innovation is undermining deterrence relationships. The fact that dozens of states are pursuing, and in some cases have already deployed, such weapons,[30] and the attendant fears of 'hypersonic arms racing',[31] appears to add fuel to the fire that these weapons mark a major change in military operations. But the nature of the challenge that these weapons pose for the global nuclear order is worth unpacking.

The first thing that needs explaining is that the label 'hypersonic missile' is slightly misleading because it refers to a speed at which other existing missiles travel—many current ballistic missiles already fly at 'hypersonic speed', that is over 3,500 miles per hour (mph).[32] The hypersonic moniker also covers two quite different military capabilities, with different attributes and potential limitations: hypersonic boost-glide vehicles (HGV) and hypersonic cruise missiles (HCMs). HGVs are warheads that are propelled into near space by a booster or rocket, before 'gliding' unpowered across the top of the atmosphere and then falling to the target. HCMs are air-breathing missiles that are powered by a scramjet engine (which is a very difficult technology to master)[33] throughout their flight inside the atmosphere. While both can travel at similar speeds, in most other senses they are quite

[30] See Richard H. Speier, George Nacouzi, Carrie A. Lee, & Richard M. Moore, 'Hypersonic missile nonproliferation: Hindering the spread of a new class of weapons', RAND Corporation, 2017.

[31] For example, R. Jeffrey Smith, 'Hypersonic missiles are unstoppable. And they're starting a new global arms race', *New York Times*, 19 June 2019, https://www.nytimes.com/2019/06/19/magazine/hypersonic-missiles.html.

[32] The Trident D5 SLBM operated by the US and UK, for example, can reach speeds of over 13,000 mph.

[33] Indeed, scramjet engines have been likened to 'keeping a candle alight in a hurricane'. For example, 'DRDO conducts scramjet engine ground test', Indian Press Release Bureau, 21 January 2025, https://www.pib.gov.in/PressReleasePage.aspx?PRID=2094886.

different: in fact, HGVs are probably more like advanced ballistic warheads (such as manoeuvrable re-entry vehicles—or MARVs),[34] and HCMs represent a faster version of current cruise missiles.

The second aspect to unpack is their stealthiness and ability to defeat current missile defence systems. HGVs—such as the Russian Avangard, the US Dark Eagle, or the Chinese DF-ZF—have been touted as game changers because they can manoeuvre to defeat interceptor missiles. But intercepting advanced ballistic missile warheads, especially when combined with various countermeasures, is already difficult. Moreover, HGVs are likely to create a considerable amount of heat as they transverse the atmosphere, making them potentially easier to track with infra-red sensors based in space, and because they slow (and cannot deploy decoys) while falling to their target, potentially easier to intercept with terminal missile defences.[35] HCMs—such as the Russian Tsirkon, the US 'Hypersonic Air Breathing Weapon Concept' (HAWC), the Chinese YJ-19, or the Russian/Indian Brahmos II—present more of a challenge for defences, but other than speed don't really provide many advantages over platforms such as the US Tomahawk, United Kingdom (UK)-French Storm Shadow, or the Russian Kalibr. HCMs are also limited in range by having to carry their own fuel for the entirety of the flight; and flying faster and lower uses more fuel.[36] This is why stealthy but slower subsonic missiles are often seen as more useful—or at least better value (discussed later in this chapter). Finally, both HGVs and HCMs would be difficult to guide via external platforms (such as a satellite) due to the plasma created by the heat during flight.[37]

This suggests that 'hypersonic missiles' are unlikely to transform nuclear deterrence in the Third Nuclear Age or provide some magic ability to carry out disarming strikes against adversaries. As Ivan Oerlich of the Federation of American Scientists explains:

> many of the claims about the revolutionary nature of hypersonic weapons do not withstand scrutiny, particularly when compared to alternatives.

[34] For more, see David Wright & Cameron Tracy, 'Hypersonic weapons: Vulnerability to missile defense and comparison to MARVs', *Science & Global Security* 31:3 (2023), pp. 68–113.

[35] James M. Acton, 'Hypersonic boost-glide weapons', *Science & Global Security* 23 (2015), p. 75.

[36] All cruise missiles are limited by a need to balance speed, weight, and altitude. In theory, for the greatest impact, cruise missiles will travel at high speeds at low altitudes—but this requires much more fuel than travelling either slower or higher.

[37] Normal cruise missiles or precision guided munitions can be very accurate as a result of different types of guidance.

> Their capabilities are not new, and today most of the missions proposed for hypersonic weapons could be accomplished cheaper and with less technical risk simply by modifying existing ballistic missiles.[38]

Or as University of California Berkeley's Andrew Reddie puts it: 'Instead of worrying that hypersonic weapons are revolutionizing warfare, or that adversaries are ahead, it is these incremental concerns that should be the focus for both scholars and policy makers moving forward'.[39] Essentially, the deployment of faster weapons will create new headaches for the conventional battlefield, not least protecting deployed troops, high-value fixed targets, and ships at sea. But at the time of writing, they don't fundamentally change the strategic equation or provide a magic bullet.

***

The concern that hackers could somehow break into the computers and networks that control nuclear weapons systems is not a distinctly Third Nuclear Age problem—it can probably be traced back to at least the 1980s as digital technologies and networks began to be incorporated into weapons and support systems in different ways, and was perhaps best highlighted by the 2010 'Stuxnet' attack on the Iranian uranium enrichment facility at Natanz.[40] In theory it may be possible to cause the direct launch of a nuclear weapon or a nuclear explosion through a computer network (or 'cyber') attack, but while the threat can never be zero, in reality this would be very difficult.

There are several reasons for this: (i) to the extent that information is publicly available, nuclear command and control systems and weapons systems are air gapped—that is physically separated from the wider internet.[41] This means that traditional hacking over a network connection is probably not possible. (ii) It is not clear why a nation state would attempt to cause the *launch* of a nuclear weapon—it seems far more likely that terrorists would seek to do this. But there are probably far easier ways for terrorist

[38] Ivan Oelrich, 'Cool your jets: Some perspective on the hyping of hypersonic weapons', *Bulletin of the Atomic Scientists* 76:1 (2020), p. 37.

[39] Andrew Reddie, 'Hypersonic missiles: Why the new "arms race" is going nowhere fast', *Bulletin of the Atomic Scientists*, 13 January 2020, https://thebulletin.org/2020/01/hypersonic-missiles-new-arms-race-going-nowhere-fast/.

[40] The best book on this is Kim Zetter, *Countdown to zero day* (New York: Crown Publishers, 2014).

[41] During research for a previous project, I was regularly told by officials that nuclear weapons were safe because they weren't connected to 'the internet'. See Andrew Futter, *Hacking the bomb: Cyber threats and nuclear weapons* (Washington, DC: Georgetown University Press, 2018).

groups to cause death and destruction than seeking to access the computers, networks, and software that control nuclear weapons. (iii) Most—though not all—nuclear weapons systems are not held on alert (the US and Russia being the outliers ...), operators use secret and encrypted communications to authorise launch, and most weapons contain permissive action links;[42] this means that it would be very difficult to cause an unauthorised launch digitally. (iv) It is conceivable that a logic bomb or back door may be introduced into the supply chain for nuclear weapons systems, but making sure this reached the correct place, was accessible, and wasn't discovered would be very hard. (v) Nuclear systems are (hopefully) amongst the best protected of all computer systems and closed networks.[43]

However, this does not mean that cyber-nuclear interactions aren't important, but rather that they exist across a broad spectrum.[44] Four risks stand out: First, an adversary with a sophisticated cyber capability seeks to prevent nuclear systems from working—this is known as 'left of launch' and is discussed below. This would still be difficult, but it is more credible to think that a state actor might seek a way to undermine the confidence of an adversary in their deterrent.[45] Second, computer network operations are employed to steal sensitive secrets—these might be about warhead or weapons designs, processes, or information about key people. This could lead to proliferation, efforts to counter those weapons, or attempts to politically embarrass or coerce the leaders of nuclear-armed states.[46] Third, increasingly digitised systems to control nuclear weapons malfunction or don't operate in the manner that they should—they aren't hacked but just go wrong because of their complexity.[47] In the First Nuclear Age nuclear

[42] A coded switch or sensor that prevents unauthorised or accidental use.

[43] Albeit, in 2013, the US Defense Science Board warned that critical information technology systems—including those used for US nuclear operations—had not been sufficiently tested to see if they would work under attack from a sophisticated opponent. 'Task force report: Resilient military systems and the advanced cyber threat', Defense Science Board, US Department of Defense, January 2013, https://www.govinfo.gov/app/details/GOVPUB-D-PURL-gpo60252.

[44] For a discussion of this, see Futter, *Hacking the bomb*; Stephen Cimbala, *Nuclear weapons in the information age* (London: Continuum, 2012); and Herb Lin, *Cyber threats and nuclear weapons* (Stanford, CA: Stanford University Press, 2021). See also Erik Gartzke & Jon R. Lindsay, 'Thermonuclear cyberwar', *Journal of Cybersecurity* 3:1 (2017), pp. 37–48.

[45] This scenario is explored in Wade L. Huntley, 'Technology threats to NC3: A future scenario', in James J. Wirtz & Jeffrey A. Larsen (eds.), *Nuclear command, control and communications: A primer on US systems and future challenges* (Washington, DC: Georgetown University Press, 2022).

[46] The release of sensitive emails during the 2016 US election is one example of this.

[47] Charles Perrow, *Normal accidents: Living with high-risk technologies* (Princeton, NJ: Princeton University Press: 1999).

command and control systems were largely analogue, today most are moving towards digitisation; this creates trade-offs.[48] And fourth, cyber capabilities are used to distort the information space, and thereby indirectly cause escalation and possible nuclear use. This is discussed in more detail in the next section.

Cyber threats to nuclear operations are something to be taken seriously, especially as nuclear infrastructures and digital ecosystems become more complex, but the greatest risk is probably not from hackers causing a launch or explosion. Notwithstanding our unending reliance on computers and networks as a species, neither is it likely that cyber attacks will be able to cause the same sort of rapid mass destruction as nuclear weapons. As Joseph Nye put it in 2011—'Destruction or discontinuation of cyber systems could return us to the economy of the 1990s—a huge loss of GDP—but a major nuclear war could return us to the Stone Age'.[49]

***

Alongside the disruptive impact of 'hypersonic' missiles and 'cyber attacks', fears of how AI might transform, complicate, and challenge existing models of stability and security have become a staple of the Third Nuclear Age technological debate.[50] Worse-case fears, such as putting computers in charge of nuclear command and control systems or the deployment of autonomous platforms armed with nuclear weapons, might sound like science fiction—and sound familiar to the plot of the 1984 film *The Terminator*—but have nevertheless become pervasive. But, as Vincent Boulanin at the Stockholm International Peace Research Institute put it in an authoritative paper on the topic:

> it is easy to misperceive the opportunities and challenges posed by AI in the military sphere. AI could enable major improvements in many areas of warfare, including the nuclear domain; however, foreseeable

[48] Andrew Futter, 'The double-edged sword: US command and control modernisation', *Bulletin of the Atomic Scientists*, 29 June 2016, https://thebulletin.org/2016/06/the-double-edged-sword-us-nuclear-command-and-control-modernization/.

[49] Joseph Nye, 'Nuclear lessons for cyber security?' *Strategic Studies Quarterly*, Winter 2011, pp. 18–38.

[50] James Johnson, *AI and the bomb: Nuclear strategy and risk in the digital age* (Oxford: Oxford University Press, 2023); Edward Geist, *Deterrence under uncertainty: Artificial intelligence and nuclear warfare* (Oxford: Oxford University Press, 2023); and Edward Geist & Andrew J. Lohn, 'How might artificial intelligence affect the risk of nuclear war', RAND Corporation, 2018, https://www.rand.org/pubs/perspectives/PE296.html.

developments will be far more prosaic than the common representation of AI in popular culture.[51]

But the fear that Alan Turing's 'thinking machines' could one day be in charge of nuclear decision making remains a major concern as we move into the Third Nuclear Age.

AI is not a single contraption, neither is it a platform or weapon. Instead, AI has been likened to electricity in the way that it might be utilised for many different purposes.[52] Broadly speaking AI software might be used to perceive, reason, learn, or act in a way that mimics human cognitive functions. Basic 'narrow' AI specialises on single tasks, while much more complex 'general' AI would be able to perform cognitive functions (e.g. think, learn, remember ...). The vast majority of what we describe as AI exists in the former category, including in the nuclear space. A good example of basic AI is an email spam filter; an example of general AI is a computer that can provide marital counselling ... This means that even if we wanted to build one today, a computer able to make decisions about nuclear war in a manner similar to humans probably doesn't exist.[53]

But this does not mean that AI won't play a role in nuclear operations, and in fact, in some ways it already does. The best example is how AI software is used to fuse and manage data from early warning systems, potentially reducing risks of false alarms.[54] AI may also be used in weapons platforms to increase accuracy or the ability to evade interception. Perhaps more worrying is that AI programmes are used in decision-making support for nuclear operations, especially during a crisis.[55] As former director of the US

[51] Vincent Boulanin (ed.), *The impact of artificial intelligence on strategic stability and nuclear risk*, Vol. I: *Euro-Atlantic Perspectives* (Stockholm: Stockholm International Peace Research Institute (SIPRI), May 2019), pp. xi–xii.

As Boulanin also notes: 'If not properly programmed or used, AI systems could misinform human decisions and actions (e.g. reinforce existing human bias or create new ones). They could also fail in unpredictable ways or be particularly vulnerable to cyberattacks. In the military context, the potential consequences of these limitations could be dramatic'. p. 3.

[52] Including playing games: In 1997, IBMs 'Deep Blue' computer defeated reigning champion Garry Kasparov in a game of chess. This appeared to mark something of an inflection point, and for some a pivotal moment in the human relationship with machines.

[53] For a discussion of this, see Bleddyn Bowen & Cameron Hunter, 'We'll never have a model of an AI major-general: Artificial intelligence, command decisions, and kitsch visions of war', *Journal of Strategic Studies* 47:1 (2024), pp. 116–46.

[54] Marion Messmer, 'What happens if AI goes nuclear?', *Chatham House*, 9 June 2025, https://www.chathamhouse.org/publications/the-world-today/2025-06/what-happens-if-ai-goes-nuclear.

[55] For a discussion of this, see Marcus Holmes & Nicholas J. Wheeler, 'The role of artificial intelligence in nuclear crisis decision making: A complement, not a substitute', *Australian Journal of International Affairs* 78:2 (2024), pp. 164–74.

Department of Defense Joint Artificial Intelligence Center, Lt General Jack Shanahan explained in 2025:

> The potential benefits of AI for national security are significant, and its diffusion rate will only accelerate. The gravest risks of AI in the nuclear enterprise, such as automation of launch authority, are well understood and relatively easy to proscribe. But other risks are more subtle and systemic. The goal is not to halt AI adoption, but to deepen understanding of its unintended consequences, especially cascading effects, emergent behaviors, and hidden influences on intelligence and decision-making.[56]

Thus, arguably the more important near-term implication of AI will be indirect in how particular software and programmes may shape the nuclear information space, particularly how policymakers and military leaders will know what information is genuine or false, and what can be relied upon and what can't (discussed below). AI will therefore principally support, or perhaps more accurately be perceived as supporting, rather than supplant decision makers, at least for the foreseeable future. Moreover, there are already various international efforts underway to create a norm that AI should not be used in nuclear command and control systems, and in 2024 US President Joe Biden and Chinese President Xi Jinping publicly agreed 'the need to maintain human control over the decision to use nuclear weapons'.[57]

***

In March 2025 US President Donald Trump announced his intention to build a 'golden dome' missile defence shield to protect the US against missile attacks.[58] It is a proposal strikingly similar to Ronald Reagan's

[56] Lt General John 'Jack' Shanahan, 'Artificial intelligence and nuclear command and control: It's even more complicated than you think', *Arms Control Today*, September 2025, https://www.armscontrol.org/act/2025-09/features/artificial-intelligence-and-nuclear-command-and-control-its-even-more.

[57] Jarrett Renshaw & Trevor Hunnicutt, 'Biden, Xi agree that humans, not AI, should control nuclear arms', Reuters, 17 November 2024, https://www.reuters.com/world/biden-xi-agreed-that-humans-not-ai-should-control-nuclear-weapons-white-house-2024-11-16/.

[58] 'Remarks by President Trump in joint address to Congress', The White House, 4 March 2025, https://www.whitehouse.gov/remarks/2025/03/remarks-by-president-trump-in-joint-address-to-congress/. The full quote is: 'As commander in chief, my focus is on building the most powerful military of the future. As a first step, I'm asking Congress to fund a state-of-the-art Golden Dome missile defense shield to protect our homeland, all made in the USA. (Applause.) And Ronald Reagan wanted to do it long ago, but the technology just wasn't there, not even close. But now we have the technology. It's incredible, actually. And other places, they have it: Israel has it. Other places have it. And the United States should have it too. Right, Tim?

announcement of the Strategic Defense Initiative over 40 years ago.[59] A key component of Trump's plan would be deploying lasers able to destroy missiles as they travel across the atmosphere or in near space. Trump's vision has fed into a broader concern about how an emerging era of competition between major powers would lead to the *weaponisation* of space. And space as a 'new' domain of operations has regularly been cited as a major flashpoint for the Third Nuclear Age.[60]

Space has essentially been *militarised* since the 1950s, as more satellites were placed in orbit to support various aspects of military operations, including the nuclear mission—initially for early warning, spying, and communications, and later to support precision targeting and warfighting. While the 1967 Outer Space Treaty prohibited the placement of nuclear weapons or other weapons of mass destruction in orbit, and banned extra-terrestrial 'hostile activities', it did not prohibit other types of weapons.[61] This includes a class of weapons known as directed energy weapons (DEW), such as gas, solid state, or chemical lasers (that use light); microwaves (that use waves); particle beams (charged particles), and electromagnetic pulse (bursts of radiation).

DEW use energy rather than a projectile/warhead, which theoretically makes them much faster, more accurate, reusable, and lower cost per shot than a kinetic weapon (such as a ballistic missile or bomb). Applications could include missile and hypersonic missile defence, anti-drone defence, 'frying' or 'blinding' satellites, radar, or other electrical systems.[62] But DEW are limited by power supply, the need for cooling, beam coherence over distance, the need to remain focussed on a presumably fast-moving target, and possibly by atmospheric conditions. This is particularly acute for plans to place lasers aboard satellites in low earth orbit for missile interception,

Right? (Applause.) They should have it too. So, I want to thank you. But it's a very important. This is a very dangerous world. We should have it. We want to be protected. And we're going to protect our citizens like never before.'

[59] And actually, to ideas that were first around in the 1950s. For a detailed overview of laser development, see Jeff Hecht, *Lasers, death rays, and the long, strange quest for the ultimate weapon* (Amherst, NY: Prometheus Books, 2019).

Although in some important ways it differs from SDI. As the former Director of Policy for the US National Nuclear Security Administration Monte Mallin points out, this is not a research and development programme, it is not linked with disarmament and doesn't seem to involve allies. See https://montemallin.substack.com/p/golden-dome-some-opening-thoughts.

[60] Indeed, this literature is now suggesting that we may be entering a 'new' space age.

[61] https://www.unoosa.org/oosa/en/ourwork/spacelaw/treaties/introouterspacetreaty.html.

[62] Henry 'Trey' Obering, 'Directed energy weapons are real ... and disruptive', *PRISM* 8:3 (2019), pp. 36–47.

where distances to targets may be significant, and satellites spend just a short period of time over a particular area.[63] However, some examples of more limited DEW programmes do exist, such as the UK Dragon Fire tested in 2024—designed for use against UAVs,[64] the US Navy's HELIOS system—designed to counter fast attack aircraft or missiles,[65] or the US airborne laser project—which for over two decades has been seen as a possible solution to 'boost-phase missile defence'.[66] Additionally, the Russian Peresvet programme has been designed for air defence and possibly anti-satellite warfare from ground-based platforms, but may have limited range as well as massive power requirements.[67] Indeed, the power requirements of laser weapons vary enormously, from a few kilowatts (e.g. the power needed by a single household for one day) for shooting down a drone to several megawatts (e.g. the power needed to drive an electric car for several thousand miles) for missile interception.[68] The main takeaway is that the idea of using DEW is not new, but applications are hard or often limited.[69]

Arming satellites with lasers is only one possible way that space could become weaponised: the concern about 'space planes', 'rods from god',[70] and the speculation that Russia may be developing a 'nuclear device to deploy in Space',[71] all play into these fears. It also technically includes the renewed interest in so-called 'FOBS' (fractional orbit bombardment system) platforms—warheads that undergo partial orbit before de-orbiting and

[63] Satellites in low earth orbit (LEO) take around 90 minutes to travel around the ~25,000 mile circumference of the earth.

[64] 'Advanced future military laser achieves UK first', Gov.uk, 19 January 2024, https://www.gov.uk/government/news/advanced-future-military-laser-achieves-uk-first.

[65] Carter Johnson, 'US Navy HELIOS laser test underscores greater advancements in Directed Energy Weapons', *Naval News*, 4 February 2025, https://www.navalnews.com/naval-news/2025/02/u-s-navy-helios-laser-test-underscores-greater-advancements-in-directed-energy-weapons/.

[66] Theresa Hitchens, 'Missile Defense Agency has new hope for airborne lasers', *Breaking Defence*, 17 June 2024, https://breakingdefense.com/2024/06/missile-defense-agency-has-new-hope-for-airborne-lasers/.

[67] Bart Hendrickx, 'Peresvet: A Russian mobile laser system to dazzle enemy satellites', *The Space Review*, 15 June 2020, https://www.thespacereview.com/article/3967/1.

[68] Dragon Fly has been rated a '50 kilowatt'. For more detail, see https://www.gov.uk/government/news/advanced-future-military-laser-achieves-uk-first.

[69] E.g. Kosta Tsipis, 'Laser weapons', *Scientific American* 245:6 (1981), pp. 51–7. For an overview, see Hecht, *Lasers, death rays*.

[70] John Aquilla, 'Rods from god: Imagine a bundle of telephone poles hurtling through space at 7000 mph', *San Francisco Chronicle*, 12 March 2006, http://www.californiaskywatch.com/warcosts/wp-content/uploads/upper-atmospheric-testing/26W_2006_Space_Based_Weapons_Taking_War_Into_Space_S.F._Chronicle_March_12_2006_Arquilla.pdf.

[71] James M. Acton, 'Space nukes are bad', Carnegie Endowment for International Peace, 8 October 2024, https://carnegieendowment.org/posts/2024/10/russia-space-nukes-bad?lang=en.

falling to targets rather than following a ballistic flight path. In 2021, China tested a FOBS armed with a hypersonic glide vehicle,[72] which could travel over the south pole to the intended targets as opposed to the shorter route (to the US / North Atlantic Treaty Organization (NATO)) over the north pole. Doing this would cause a problem for north-facing radar and early warning systems, and mean constructing different early warning capabilities, but could probably be countered. Moreover, as the Soviet Union found out in the 1970s, sacrifices in payload and accuracy must be made in a FOBS system *vis-à-vis* a ballistic missile. Ultimately, while space as a domain has become more important to nuclear operations, weaponisation through space-based lasers and other counterspace weapons have major limitations.

***

Lastly, and arguably because it is the most futuristic is Quantum. At the time of writing, perhaps the best thing to say about the possible impact of 'quantum' technologies is that they *could* have considerable impact on nuclear operations, but that it really is too early to tell. Quantum refers to the behaviour of particles and energy at the subatomic level and challenges the traditional understanding of physics. To give some idea, if an atom (approximately 100,000 times thinner than a human hair) were the size of a normal sports stadium, then a quark (which forms protons and neutrons) would be smaller than a grain of soil. At this scale, the science can be quite difficult to get your head around, but it is basically the theory that matter can exist in multiple linked forms at the same time. One way of thinking about it is through a thought experiment proposed by Austrian physicist Edwin Schrodinger, whereby a cat is placed in a sealed box with something that can kill it, but it will only die if you open the box. But because you cannot observe it, it is both alive and dead at the same time: it can exist in multiple states.[73] Technically, as the physicist Michal Krelina points out, the 'first quantum revolution' gave us nuclear energy, semiconductors, and lasers, what we are now in is the 'second quantum revolution'. This revolution:

> does not bring fundamentally new weapons or standalone military systems, but rather significantly enhances measurement capability, sensing,

[72] Bleddyn Bowen & Cameron Hunter, 'Chinese fractional orbital bombardment', *APLN Policy Brief* 78 (2021), https://cms.apln.network/wp-content/uploads/2021/11/FINALBowenHunterPolicyBrief.pdf.

[73] Joshua Howgego, 'Schrodinger's cat', *New Scientist* (nd), https://www.newscientist.com/definition/schrodingers-cat/.

precision and computation power and efficiency of the current and future military technology.[74]

Essentially, operating at the quantum level allows for dramatically improved performance in certain domains over 'classical' computers. Five *possible* applications for nuclear operations stand out. First, quantum computers could be used to break encrypted communications (such as leadership communications and perhaps even launch orders). Second, quantum sensors such as gravimeters, atomic clocks, magnetometers, radiation detectors, and radar could significantly enhance timing, sensing, and imaging capabilities, and be used to 'find' previously stealthy platforms (such as nuclear-armed aircraft or submarines). Third, quantum communications through superimposed qubits could replace hackable bits and digital keys and could thereby provide the possibility of secure communications and verification (through, for example, quantum key distribution). Fourth, 'quantum AI' could theoretically provide enhanced decision-making support or data processing.[75] Fifth, quantum could be used as an 'unjammable' replacement for global positioning systems (GPS).[76] At the time of writing, IBM's 'Quantum Osprey'[77] processor and Atom Computing[78] are examples of the technology, and there are also current promising uses in biology. But quantum still remains largely experimental.

***

It is easy to assume the worst when it comes to technological change and fall into the trap of either thinking that anything that might be possible will happen, or to dismiss such possibilities as science fiction. As with most things, the reality is somewhere in middle. Very fast manoeuvrable missiles

[74] Michal Krelina, 'Quantum technology for military applications', *EPJ Quantum Technology* 8:24 (2021).

[75] Michael van Amerongen, 'Quantum technologies in defence and security', *NATO Review*, 3 June 2021, https://www.nato.int/docu/review/articles/2021/06/03/quantum-technologies-in-defence-security/index.html.

[76] Madeleine North, 'What is quantum navigation—and could it replace GPS?', World Economic Forum, 5 July 2024, https://www.weforum.org/stories/2024/07/what-is-quantum-navigation-earth-observation/.

[77] Hugh Collins & Chris Nay, 'IBM unveils 400 qubit-plus quantum processor and next generation IBM quantum system two', IBM, 9 November 2022, https://newsroom.ibm.com/2022-11-09-IBM-Unveils-400-Qubit-Plus-Quantum-Processor-and-Next-Generation-IBM-Quantum-System-Two.

[78] Alex Wilkins, 'Record-breaking quantum computer has more than 1000 qubits', *New Scientist*, 24 October 2023, https://www.newscientist.com/article/2399246-record-breaking-quantum-computer-has-more-than-1000-qubits/.

with nuclear or conventional warheads could create headaches for defence planners in certain circumstances—but not necessarily in a completely different way to those that have been available for some time; 'cyber' threats are diverse and much more acute at the lower end of the threat spectrum—targeted attacks for strategic effect are difficult, time-consuming, and often quite limited in their impact; the relationship between humans and thinking machines is evolving, but both prudence and technology will likely limit the direct impact on nuclear operations, and hopefully no one has an interest in fully automated nuclear command and control systems; DEW may offer different ways of pursuing established missions but have limitations; and the applications of quantum seem mostly likely something for the longer term rather than the breakthrough of tomorrow.

A quiet metamorphosis: Smarter, real-time, dispersed, and more interlinked technologies

While much focus has been given to the exotic, sexy, and ostensibly game-changing technologies described above, the really significant development has been more gradual and evolutionary. But this is not to say that it isn't important, or that the impact will not be widespread, rather that it is more nuanced than often reported. In general, change is being driven by advances in enabling systems such as sensors, computer processing, and data management and in support infrastructure technology as much as by 'front-end' 'effector' capabilities.

Perhaps the most significant technological dynamic of the Third Nuclear Age is the development, enhancement, and spread of precision missiles that might be used to deliver either nuclear or conventional payloads. These range from relatively slow *subsonic* (500 mph—the speed of a passenger airliner) to slightly faster *supersonic* (over 900 mph) cruise missiles, to much faster ballistic missiles of different ranges. As noted earlier, US scholar Dennis Gormely had warned about the impact of the spread of cruise missile technology in his book *Missile Contagion* back in 2010,[79] and especially in the US, different missiles have been becoming increasingly accurate since the 1980s. But the difference in the Third Nuclear Age is that the technology is spreading, and accuracy is reaching a point whereby long-range precision strikes against high-value (possibly nuclear or nuclear-related) targets

[79] Gormley, *Missile contagion.*

is becoming a possibility with conventional rather than nuclear weapons. Moreover, some experts believe that it is increasingly possible to destroy high-value *mobile* targets too.[80]

This is significant because during the First Nuclear Age any 'counterforce' attack (consisting of leadership and military targets, including nuclear platforms) against an opponent could only really be undertaken with nuclear weapons, and during the Second Nuclear Age it was only really the US that began thinking about precision weapons to target limited nuclear threats.[81] But today, all nuclear-armed states, and many non-nuclear-armed states have developed—or are seeking to develop—the ability to hit targets at distance with high accuracy, often with enhanced stealth. HCMs and HGVs fall into this category but should only be seen as one component of the wider missile challenge.

Precision missiles have been enabled by two critical developments: advances in the ability to locate and, if necessary, track targets, especially using space-based sensors, and advances in guidance systems. Advances in guidance systems include the use of external support from satellites or other global positioning and navigation platforms as well as enhanced software inside the missile such as TERCOM (terrain contour matching) or DSMAC (digital scene matching area correlation) as well as datalinks and inertial guidance (using the stars). This has allowed accuracy down to a matter of metres (known as circular error probable—CEP) over hundreds of miles. While it is difficult to find official sources, the latest US Tomahawk cruise missile is believed to be accurate to around 5 metres up to 1,000 miles,[82] the UK-French Storm Shadow approximately 3 metres over 350 miles,[83] the US JASSM around 3 metres over 250 miles,[84] the Russian Iskander around 5–10 metres over several hundred miles,[85] and the intercontinental-range

[80] E.g. Paul Bracken, 'The hunt for mobile missiles: Nuclear weapons, AI and the new arms race', Foreign Policy Research Institute, September 2020, https://www.fpri.org/wp-content/uploads/2020/09/the-hunt-for-mobile-missiles.pdf.

[81] The pros and cons of this phenomenon are captured expertly by the Carnegie Endowment's James Acton in his 2013 monograph *Silver bullet*.

[82] 'Tomahawk cruise missile', *Britannica*, updated 27 April 2025, https://www.britannica.com/technology/Tomahawk-cruise-missile.

[83] Boyko Nikolov, 'UK's Storm Shadow becomes an "iron block" without US intel', BulgarianMilitary.com, 6 March 2025, https://bulgarianmilitary.com/2025/03/06/uks-storm-shadow-becomes-an-iron-block-without-u-s-intel/#:~:text=In%20its%20final%20approach%2C%20an,battlefield%20changes%20is%20severely%20compromised.

[84] 'JASSM / JASS ER', CSIS Missile Threat Project, updated 23 April 2023, https://missilethreat.csis.org/missile/jassm/#easy-footnote-bottom-10-838

[85] '9K720 Iskander (SS-26)', CSIS Missile Threat Project, updated 23 April 2023, https://missilethreat.csis.org/missile/ss-26-2/.

Trident missile (operated by the US and UK) ~90 metres over thousands of miles.[86] The Chinese DF21 'ship killer' may be accurate to around 20 metres over distances beyond 1,000 miles.[87] By comparison, early ballistic and cruise missiles (Minuteman I or Pershing II, for example) were believed to be accurate to around 1,000 metres / 1 kilometre.[88] This is the difference between a warhead being very likely to land in the front room of your house, versus being likely to land somewhere in your village/town. As Keir Lieber and Daryl Press put it:

> The accuracy revolution is ongoing. As accuracy continues to improve, the effectiveness of conventional attacks on hard targets will continue to increase. Today, low-yield nuclear weapons can destroy targets that once required very large-yield detonations. In the future, many of those targets will be vulnerable to conventional attacks.[89]

Indeed, the Third Nuclear Age missile challenge also includes the possibility of using current ballistic missiles for conventional roles: in 2006 the US government considered deploying a 'conventional trident' missile;[90] in 2024, Russia was believed to have launched an intermediate-range conventionally armed 'Oreshnik' ballistic missile attack against Ukraine,[91] and the Pakistani expert Rabia Akhtar has described the Pahalgam Crisis of 2025 as the 'first missile war of the Third Nuclear Age'.[92] It is even possible that we see a renewed interest in nuclear-powered missiles, although in August 2019 a suspected test of the Russian 'Burevestnik' nuclear-powered cruise missile resulted in a radiation accident.[93] The fact that ballistic missile technology is very similar to that used for space launch vehicles—a capability

[86] See 'Facts about Trident', Nuclear Information Service, January 2008, https://www.nuclearinfo.org/wp-content/uploads/2020/09/Facts-about-Trident.pdf.

[87] Peter Suciu, 'DF-21D and DF-26B: China's missiles build to sink navy aircraft carriers', *The National Interest*, 15 August 2024, https://nationalinterest.org/blog/buzz/df-21d-and-df-26b-chinas-missiles-built-sink-navy-aircraft-carriers-207308.

[88] 'Minuteman I', CSIS Missile Threat Project, updated 23 April 2023, https://missilethreat.csis.org/missile/minuteman-i/.

[89] Liber & Press, 'The myth of the nuclear revolution, p. 80.

[90] This idea also made it across the Atlantic to the UK: Robert Green, 'Conventionally-armed UK trident?' *The RUSI Journal* 147:1 (2002), pp. 31–4.

[91] Sidharth Kaushal & Matthew Savill, 'The Oreshnik ballistic missile: From Russia with love?', *RUSI Commentary*, 10 December 2024, https://www.rusi.org/explore-our-research/publications/commentary/oreshnik-ballistic-missile-russia-love.

[92] In a conversation with the author.

[93] Ankit Panda, 'Russia conducts test of nuclear-powered cruise missile', *The Diplomat*, 6 February 2019, https://thediplomat.com/2019/02/russia-conducts-test-of-nuclear-powered-cruise-missile/.

being pursued by many states—is only likely to add to the diffusion of missile latency.

A second major technological development that will shape the Third Nuclear Age is the development and spread of missile defence systems. Interest in defence against cruise and ballistic missiles can be traced back to the Second World War as Nazi V1 and V2 Vergeltungswaffe or 'vengeance weapons' began falling on London. The first defensive systems began being deployed in the 1950s—such as the US Nike-Ajax and Safeguard, the Russian Galosh/A35, and the lesser known British Violet Friend[94]—and all of these involved using interceptor missiles armed with nuclear weapons to destroy the incoming warhead in the blast.[95] For most of the Cold War years ballistic missile defence was dominated by the two superpowers, albeit limited after 1972 by the ABM Treaty. Several decades later the technology has improved markedly, and such systems are now being deployed by numerous different actors for different purposes.

The missile defence mission is not homogenous, and well-publicised debates about whether missile defence 'works or not' rather miss the point.[96] Some missile defence systems deployed to counter certain threats are better than others. There is quite a marked difference, for example, between trying to shoot down short-range conventionally armed rockets fired across a border with multiple proximity explosions, and intercepting numerous nuclear warheads armed with countermeasures travelling through space at thousands of miles an hour by colliding directly with them kinetically or with directed energy. While the speed and difficulty of the interception mission varies considerably, most missile defence systems involve the following process: sensors to spot the launch; radar or satellites to track the incoming warhead; processors to calculate the trajectory, speed, and likely target of the warhead; a method of interception—such as a missile, rocket, or possibly a laser or other form of directed energy—and a radar to help guide the interceptor to the target.

Today missile defences are far from fool-proof, and testing records remain mixed. The US ground-based mid-course defence system, for example, has

[94] Much less well known than the US and Soviet programmes. See Jeremy Stocker, *Britain and missile defence: 1942–2022* (London: Frank Cass, 2004).

[95] This would still involve a nuclear explosion in near space, but this was deemed preferable to allowing the warhead to explode above a target.

[96] A good example being the debate about the Patriot missile defence system during the 1991 Gulf War, and how this was used by some to justify deploying much more complex systems. See Postol, 'Lessons of the Gulf War, pp. 119–71.

succeeded in 12 out of 21 tests since 1999,[97] while the Patriot short-range missile defence system used by Ukraine in the war against Russia has been very successful, and it is believed that Israel successfully intercepted the majority of several hundred Iranian drones, and cruise and ballistic missiles launched during two attacks in 2024.[98] Nevertheless, the overall trajectory over the past two decades has been for missile defence systems to play an increasingly important role in the deterrence and conventional warfighting thinking of many states. At the strategic nuclear level this matters because it creates the possibility of undermining an adversary's ability to threaten retaliation (by intercepting their nuclear-armed missiles) and therefore strategic stability. At the substrategic level, it creates new challenges for the use of missiles to hit high-value targets during a conflict. Historically, the deployment of missile defences has been seen as a driver of missile proliferation as those impacted sought the ability to overwhelm the defences of an adversary.

So far, the missile defence mission has primarily focussed on what might be termed 'right of launch' capabilities—that is seeking to shoot down a missile or warhead *after* it has been launched (including hypersonic defence).[99] But while most of the above is incremental and evolutionary (though still significant), the fact that anti-missile defence now also increasingly includes efforts to prevent missiles being launched, known as 'left of launch', through non-kinetic means, is potentially more transformative.[100] While in theory it is possible to destroy a missile before launch through a pre-emptive strike with a missile, bomb or perhaps a special forces operation, attention has now shifted to achieving the same outcome with computer network operations (or hacking/cyber) and electronic measures.[101] This fundamentally shifts the possible impact of the missile defence mission.

[97] See 'Ballistic missile defense intercept flight test record', US Missile Dense Advocacy Alliance, March 2023, https://missiledefenseadvocacy.org/missile-defense-systems-2/missile-defense-intercept-test-record/u-s-missile-defense-intercept-test-record/. Interestingly, the US Missile Defense Agency no longer publishes the test record.

[98] Jen Judson, 'How Patriot proved itself in Ukraine and secured a fresh future', *Defense News*, 9 April 2024, https://www.defensenews.com/land/2024/04/09/how-patriot-proved-itself-in-ukraine-and-secured-a-fresh-future/.

[99] For example, Tom Karako & Masao Dahlgren, 'Complex air defense: Countering the hypersonic missile threat', Center for Strategic and International Studies Missile Defense Project Report, February 2022, https://www.csis.org/analysis/complex-air-defense-countering-hypersonic-missile-threat.

[100] See 'Trump inherits secret cyber against North Korean missiles', *New York Times*, 4 March 2017, https://www.nytimes.com/2017/03/04/world/asia/north-korea-missile-program-sabotage.html.

[101] See, for example, Caroline Houck, 'Left of launch missile defense: "You don't want to have just one solution to the threat"', Defense One, 24 January 2018, https://www.defenseone.com/threats/2018/01/left-launch-missile-defense-you-dont-want-have-just-one-solution-

The third key component of the quiet technological metamorphosis of the Third Nuclear Age is the spread and sophistication of different sensors, platforms, and support systems. Much like with different missiles and missile defence, many of these technologies are not new, but appear to be finding much wider expression and applicability than before. Three parts to this stand out: the use of satellites to locate and track targets on land, in the sea, in the air, or travelling through space; the ability to deploy systems underwater that might be used to 'find' and compromise submarines, including those armed with nuclear weapons, and; the ability of radar and other air defence systems to identify and shoot down stealthy aircraft. Each of these is perhaps better thought of as an ongoing competition, or to use the Cold War jargon, the latest iteration of 'nuclear hide and seek'. What cuts across each domain is the increasing development of various types of autonomous platforms to support these missions.

Ever since the launch of Sputnik there have been concerns that satellites could be used to find strategic assets and provide the information required to target them. Missile fields located far away from borders and prying eyes suddenly became more vulnerable (high-altitude reconnaissance aircraft were already vulnerable to air defences—as US U2 pilot Gary Powers found out in 1960), but while early warning satellites could pick up launches through their heat signatures, they had little capacity to support interception or precision strikes. During the First Nuclear Age it was therefore only really *fixed* targets that became vulnerable to *nuclear* blasts—largely as a result of warhead accuracy, and orbital platforms could track but not easily be used to intercept projectiles travelling through space. As discussed above, this is why early missile defence systems relied on a nuclear interception. As a result, mobile missiles were considered to be largely invulnerable because they could move and hide.

But what seems to have changed in the Second and now Third Nuclear Age is that more satellites have been placed in orbit with ever-better sensors, and the associated processing capabilities back on earth. Potentially, these might not just allow for smaller nuclear or conventional attacks on fixed targets, but also theoretically to track and hold vulnerable mobile platforms too. This is especially the case when combined with other sources of intelligence—or 'high-touch' reconnaissance strategies involving multiple

threat/145438/:~:text=Lawmakers%20are%20pushing%20boost-phase,missiles%20before %20they%20lift%20off.

platforms as Paul Bracken puts it.[102] While privately owned satellite constellations such as 'Starlink' may not currently be used directly for targeting, they have become very useful for the rapid analysis and communication of targeting *data* in Ukraine.[103]

At the same time, these space-based sensors have become better at tracking warheads or other platforms travelling through low earth orbit, which when linked with ground-based radar and data fusion apparatus make intercepting missiles increasingly possible. Of course, these assets in space are also potentially vulnerable to interference or direct attack, and in the past decade we have seen the return of different types of anti-satellite weapons.[104] In 2007 China successfully destroyed a weather satellite; in 2008 the US shot down a malfunctioning reconnaissance satellite, in 2019 India destroyed the Microsat-R in low earth orbit, and in 2021, Russia destroyed its own 'Kosmos 1408' satellite—sending hundreds of pieces of debris into low earth orbit. Despite the 2021 announcement that the US would halt the testing of direct-ascent ASAT weapons, and the possibility of a 'Use of Force against outer space objects treaty', the dual use nature of many missile and missile defence capabilities makes this difficult to enforce in practice.[105]

In a similar vein, nuclear-powered submarines hidden in the ocean were also believed largely impervious to attack due to the size of the ocean, their capacity for long patrols, and their comparative quietness. While research has emerged that First Nuclear Age US/NATO anti-submarine warfare capabilities may have been better than widely realised,[106] finding and destroying nuclear powered submarines armed with nuclear tipped ballistic missiles (ship submersible ballistic nuclear, or SSBNs) was considered a difficult task, and therefore so-called 'boomers' formed the basis of a secure nuclear

[102] Paul Bracken, *The hunt for mobile missiles: Nuclear weapons, AI and the new arms race* (Philadelphia, PA: Foreign Policy Research Institute, 2020), pp. 6–7.

[103] An additional challenge posed by Space X, Blue Origin, and others is the increasing role of non-state actors in space. As of 2025, nearly half of all satellites were owned by non-government corporations.

[104] See Brian Weeden & Victoria Samson, 'Global counterspace capabilities: An open assessment', Secure World Foundation, April 2020, https://swfound.org/media/206957/swf_global_counterspace_april2020_es.pdf. For an overview, see Greco, 'A history of anti-satellite programs'.

[105] Ankit Panda & Benjamin Silverstein, 'The US moratorium on anti-satellite missile tests is a welcome shift in space policy', Carnegie Endowment for International Peace, 20 April 2022, https://carnegieendowment.org/posts/2022/04/the-us-moratorium-on-anti-satellite-missile-tests-is-a-welcome-shift-in-space-policy?lang=en.

[106] Austin Long & Brendan Rittenhouse Green, 'Stalking the secure second strike: Intelligence, counterforce, and nuclear strategy', *Journal of Strategic Studies* 38:1–2 (2015), pp. 38–73.

second strike. But in the Third Nuclear Age, technological advances are potentially making it harder to hide submarines. As the naval warfare expert Bryan Clark points out ...

> Since the Cold War, submarines—particularly quiet American ones—have been assumed to be largely immune to anti-access threats. Yet the ability of submarines to hide through quieting alone will decrease as each successive decibel of noise reduction becomes exponentially more expensive and new detection techniques mature that rely on phenomena other than the sounds emanating from a submarine.[107]

These techniques include active and passive acoustic sensors, such as sonar or towed arrays; magnetic anomaly detectors; satellites and overhead surveillance, synthetic aperture or infrared radar; or hydrophones, such as the US sound surveillance system (SOSUS),[108] all using AI-enabled software to distinguish from background interference. A key component of this may increasingly be the use of uninhabited underwater vehicles (or UUVs) able to loiter, search and track submarines, or—as with the Russian Poseidon Status-6 torpedo, potentially deliver nuclear payloads too.[109] However, as former director of the MIT Security Studies Program Owen Cote notes:

> the survivability of nuclear-powered and nuclear-armed ballistic missile submarines, also known as SSBNs, has and will continue to vary greatly, depending upon whose SSBNs one is assessing, who might be looking for them, the state of pro- submarine and anti-submarine technology deployed by the hider and the searcher, and the maritime geography that shapes and channels their competition.[110]

[107] Clark continues: 'While the physics behind most of these alternative techniques has been known for decades, they have not been exploitable until very recently because computer processors were too slow to run the detailed models needed to see small changes in the environment caused by a quiet submarine'. Bryan Clark, 'The emerging era in undersea warfare', Center for Strategic and Budgetary Assessments, 2015, p. 8.

[108] See https://www.usni.org/magazines/naval-history-magazine/2021/february/66-years-undersea-surveillance.

[109] Silky Kaur, 'One nuclear-armed Poseidon torpedo could decimate a coastal city. Russia wants 30 of them', *Bulletin of the Atomic Scientists*, 14 June 2014, https://thebulletin.org/2023/06/one-nuclear-armed-poseidon-torpedo-could-decimate-a-coastal-city-russia-wants-30-of-them/.

[110] Owen R. Cote, 'Invisible nuclear-armed submarines, or transparent oceans? Are ballistic missile submarines still the best deterrent for the United States?', *Bulletin of the Atomic Scientists* 75:1 (2019), pp. 30–5.

The general feeling is that the US and UK are (currently) better at this than others. Nevertheless, submarines remain the go-to platform for a secure second nuclear strike, and more states are moving towards acquiring ultra-stealthy underwater nuclear-powered systems (notably Australia as part of the AUKUS (Australia-UK-US treaty) agreement and Brazil with its indigenous 'Prosub' programme).[111] Nuclear propulsion remains the gold standard, but several states are developing very advanced air-independent propulsion (AIP) and other diesel-electric platforms too.[112] AIP works by using a closed cycle engine or fuel cells; diesel subs rely on charging batteries.

Lastly, in the air we are seeing the latest iteration between increasingly stealthy aircraft, their spread to more actors, and progressively capable air defences. For most of the First and Second Nuclear Age, stealth aircraft was the preserve of the US—notably the F117 'nighthawk' fighter jet and B2 'spirit' strategic bomber, but so-called '5th generation' aircraft are now being developed by other states too. The big difference between 4th and 5th generation aircraft is that stealth is built in from the start. This includes use of radar-absorbent materials, radar reflection body design, internal weapons bays, low heat signatures, and minimal radio emissions. Current examples include the US B21 bomber, F22 and F35 fighters, as well as the Russian SU-57 and Chinese J-20 fighter jets—all of which could be used for nuclear missions.[113] By way of an example, the cross-section of the B2 spirit has been likened to 'somewhere between a small bird and a marble', while the B21 may be 'smaller than an insect'.[114] Stealth aircraft are significant because they provide a way of carrying out unobserved attacks—potentially with tactical nuclear weapons—rather than having to rely on stand-off capabilities like cruise missiles. They also potentially offer a way of penetrating A2AD. The US strikes against Iran in June 2025 are a good example of this.

[111] See https://www.asa.gov.au/aukus; and 'Brazil is developing nuclear-powered submarine', Government of Brazil, 24 June 2022, https://www.gov.br/en/government-of-brazil/latest-news/2022/brazil-is-developing-nuclear-powered-submarine.

[112] See Ahmad Ibrahim, 'Air-independent propulsion (AIP) submarines (SSP): Bridging the gap between conventional (SSK) and nuclear powered attack submarines (SSN)', *Defense Talks*, 26 December 2024, https://defensetalks.com/air-independent-propulsion-aip-submarines-ssp-bridging-the-gap-between-conventional-ssk-and-nuclear-powered-attack-submarines-ssn/.

[113] The F22 is currently not certified for nuclear missions.

[114] 'Radar cross section', Globalsecurity.org, n.d., https://www.globalsecurity.org/military/world/stealth-aircraft-rcs.htm; and Eric Tegler, 'A paper backing the B-21 claims its smaller than an insect to radar', Forbes, 26 December 2023, https://www.forbes.com/sites/erictegler/2023/12/26/a-paper-backing-the-b-21-claims-its-smaller-than-an-insect-to-radar/.

Uninhabited aerial vehicles (UAVs) are also likely to play a greater role in missions with strategic effects, ranging from sensors and shooters used for air and missile defence, advanced intelligence, surveillance, and reconnaissance (ISR), as well as potentially to carry out targeted attacks.[115] But this doesn't necessarily mean that we will see an end to the utility of 'big platforms' in favour of cheap expendable munitions as some have suggested. Rather we are likely to see combinations of expensive, sophisticated, cutting-edge systems alongside and networked with cheaper, simpler, and more replaceable ones.[116] Future stealth aircraft or nuclear-powered submarines controlling their own swarm of drones might be one example of this.[117]

It is important to think of the technological challenge of the Third Nuclear Age as not just about particular platforms and systems, but about a shifting information environment or ecosystem within which nuclear strategy and nuclear politics play out. At its heart is a move away from an analogue world where information was often controlled and slow to spread, to a digital world where masses of information are available to many more actors and in real time. One way to conceptualise this is to consider how the Cuban missile crisis of 1962 might play out today. As Danielle Jablanski, Herb Lin, and Harold Trinkunas explained in a 2020 monograph:

> During the Cuban Missile Crisis, analog systems, government monopoly on both secret and public information, and the relatively lengthy decision time available to leaders helped to avoid disaster ... if today's media environment existed during the Cuban Missile Crisis, American and possibly Soviet leaders would have been awash in a sea of unverified or unverifiable, emotionally laden, and politically fraught information on their adversary's intentions, force postures, and public opinion.[118]

[115] Zachary Kallenborn & Philipp Bleek, 'Swarming destruction: Drone swarms and chemical, biological, radiological and nuclear weapons', *The Nonproliferation Review* 25:5–6 (2018), pp. 523–43.

[116] E.g. Michael C. Horowitz, Lauren A. Kahn, & Joshua Schwartz, 'What drones can—and cannot—do on the battlefield', *Foreign Affairs*, 4 July 2025, https://www.foreignaffairs.com/united-states/what-drones-can-and-cannot-do-battlefield.

[117] See Alex Holdings, 'America's next stealth fighter will rule the skies', *Popular Mechanics*, 8 August 2023, https://www.popularmechanics.com/military/aviation/a44589841/stealth-fighter-jets/.

[118] Danielle Jablanski, Herbert S. Lin, and Harold A. Trinkunas, 'Retweets to midnight: Assessing the effects of the information ecosystem on crisis decision making between nuclear weapons states', in Harold A. Trinkunas, Herbert Lin, Benjamin Loehrke (eds.), *Three tweets to midnight: Effects of the global information ecosystem on the risk of nuclear conflict* (Stanford, CA: Hoover Institution Press, 2020), pp. 12–13, https://www.hoover.org/research/three-tweets-midnight.

Essentially, we now live in a world where everything is real time, connected, and in theory more susceptible to interference. Perhaps even more than this, some have argued that we increasingly live in a 'post-truth' world ...[119]

The use of various social media platforms is a key component of the Third Nuclear Age ecosystem, and especially the ability for leaders to provide knee-jerk comments on unfolding events that everyone can see.[120] Not only does this make understanding official policy more difficult, but it also opens up the potential for misunderstanding, and possibly interference. As former director of the US Defense Threat Reduction Agency, Rebecca Hersman put it (at the start of the 2020 Covid pandemic): 'today's global pandemic crisis offers a glimpse into how a toxic mix of disinformation, conspiracy theories, and digital technology can complicate effective crisis management, fuel competition and rivalry, shift blame, and sow mistrust'. In part because 'Escalation is less predictable and less linear, and includes disinformation and weaponized social media'.[121]

Technological change in the Third Nuclear Age will therefore largely be evolutionary, and about combinations of technologies rather than single specific systems creating revolutionary 'ruptures'. As the next chapter explains, technological change—especially the spread of increasingly precise missiles, better air and missile defences, more sophisticated and networked platforms and sensors, as well as new ways to avoid detection—are already challenging established notions of deterrence, driving forward the modernisation of nuclear systems and opening up new possibilities for managing nuclear threats. But perhaps most importantly, the Third Nuclear Age will be a globalised, digital, networked age, where the context of nuclear politics has shifted markedly from what has gone before.

Conclusion: Nuclear techno-politics

Douglas Adams (the author of the *Hitchhikers Guide to the Galaxy*) once famously declared that 'we are stuck with technology when what we really

[119] E.g. Andrew Facini, 'Post-truth disasters: We're not ready for nuclear war on social media', *Bulletin of the Atomic Scientists*, 19 November 2025, https://thebulletin.org/2025/11/post-truth-disasters-were-not-ready-for-nuclear-war-on-social-media/.

[120] See Trinkunas et al., *Three tweets to midnight*; and Heather Williams & Alexi Drew, 'Escalation by tweet: Managing the new nuclear diplomacy', Kings College London, July 2020, https://www.kcl.ac.uk/csss/assets/escalation-by-tweet-managing-the-new-nuclear-diplomacy-2020.pdf.

[121] Rebecca Hersman, 'Wormhole escalation in the new nuclear age', *Texas National Security Review*, Summer 2020, https://tnsr.org/2020/07/wormhole-escalation-in-the-new-nuclear-age/.

want is just stuff that works.'[122] And in some ways this captures the nature of technological change at the heart of the Third Nuclear Age: while AI, cyber, hypersonics, non-kinetic space weapons, quantum, and other exciting but equally disquieting technologies grab the headlines, the real challenge is being posed by the spread of established technologies such as ballistic and cruise missiles, stealthy aircraft and underwater platforms, better air and missile defences, as well as increasingly capable sensors in different domains and the mechanisms to fuse, share, and utilise this data rapidly. While the US remains at the forefront of most of these technologies, and retains a certain qualitative edge, the gap may be shrinking, and technological diffusion is far more widespread than in the previous two nuclear ages.

Technological change has always been a key feature of the nuclear story, often but not always accompanied by a certain technological determinism, hype, and worst-case scenario planning, and there have undoubtedly been periods of destabilising and worrying technological transformation in the past.[123] In fact, many of the technologies of interest or concern today were conceived (largely in the US) in the First Nuclear Age, began development to meet perceived requirements in the Second Nuclear Age, but only now have reached fruition and have spread in the Third Nuclear Age. This highlights the point that technological evolution doesn't happen in a vacuum or impact all actors in the same way and at the same time.

But the current era seems to be different in three important respects: first, it is not one weapon system but a whole suite of capabilities that are emerging often in parallel that can act as force multipliers; second, these systems don't just involve nuclear weapons, but increasingly *non-nuclear* platforms that might have strategic effects; and third, this is a global phenomenon with many more players and interactions than before. This is not to say that change is necessarily revolutionary, but that it does drive the need for a deeper multifaceted analysis.

As noted in the introduction to this chapter, one of the main reasons for engaging seriously with theories and realities of technological change[124] in the nuclear realm is because it is a fundamental component of what has

[122] Quoted in Jonathan Becher, 'Douglas Adams' technology rules', Forbes, 7 July 2014, https://www.forbes.com/sites/sap/2014/07/07/douglas-adams-technology-rules/.

[123] For example, B.T. Feld, T. Greenwood, G.W. Rathjens, & S. Weinberg (eds.), *Impact of new technologies on the arms race* (Cambridge, MA: MIT Press: 1971); J.J. Gertler, *Emerging technologies in the strategic arena: A primer* (Santa Monica, CA: RAND Corporation, March 1987); and Carl Builder, *Strategic conflict without nuclear weapons* (Santa Monica, CA: RAND Corporation, April 1983).

[124] See Daniel McCarthy, *Technology and world politics: An introduction* (London: Routledge, 2017).

become known as the security dilemma. The security dilemma refers to situations where actions taken by one actor to enhance their security (such as developing a new weapons system) makes others less secure, often prompting them to develop similar or counteracting technologies. The result is that both often feel less secure, arms racing can be fuelled, confusion and fear can be exacerbated, and new pathways to unintended escalation can be created.[125]

[125] See Nicholas J. Wheeler & Ken Booth, *The security dilemma: Fear, cooperation and trust in world politics* (London: Palgrave MacMillan, 2008).

3

Rethinking the unthinkable

Introduction: Nuclear deterrence works ... until the day it doesn't

The value of nuclear deterrence as both an analytical concept and real-world phenomenon has always been the source of fundamental debate, and this will be no different in the Third Nuclear Age.[1] For some, the deterrent effect of nuclear weapons is the main reason why we have not seen major war between nuclear-armed states since 1945 (or more correctly 1949, when the United States' (US) nuclear monopoly was broken)—and notably why the Cold War didn't turn hot and more recent conflicts in Ukraine and south Asia haven't escalated further. Under this logic the use of nuclear weapons is so terrible that no sane leader would take the risk of initiating war against a nuclear-armed opponent. Thus, stability with nuclear weapons is possible, and—for an albeit smaller cadre—even future nuclear proliferation is not necessarily a disaster.[2] For others, there isn't enough (or any) evidence to suggest that nuclear weapons are the main reason why major interstate war hasn't happened between nuclear-armed powers, and point to that the fact that the past should not be used as any prediction for the future. Under this logic, the risks of miscalculation, unintended escalation, accidents, institutional cockups, and unpredictable leaders mean that so long as nuclear weapons exist there is a chance they will be used, and proliferation to more actors simply increases this threat. This is of course a rather oversimplified characterisation of the deterrence debate: different geographical and historical contexts, different actors and leaders, different types of weapons systems, all shape the nature of nuclear deterrence and is its likelihood of working or

[1] The classic exposition of which is Scott Sagan & Kenneth Waltz, *The spread of nuclear weapons: An enduring debate* (New York: W.W. Norton, 2012).

[2] E.g. Kenneth Waltz, 'Why Iran should get the bomb: Nuclear balancing would mean stability', *Foreign Affairs* 91:4 (2012), pp. 2–5.

The Third Nuclear Age. Andrew Futter, Oxford University Press. © Andrew Futter (2026).
DOI: 10.1093/9780198944126.003.0004

failing. As Vipin Narang and Scott Sagan explain in the introduction to their excellent 2022 edited book *The Fragile Balance of Terror*:

> The non-use of nuclear weapons since Hiroshima and Nagasaki in 1945 has generated significant overconfidence and complacency in the restraining and stabilizing effects of nuclear weapons and deterrence. We do not deny that nuclear weapons can induce, and have induced, caution among their possessors ... But that powerful deterrent is by no means a perfect deterrent.[3]

Today, in stark contrast to the disarmament optimism inherent in the early Second Nuclear Age, where cuts in nuclear stockpiles were made by the US, Russia, France, and the United Kingdom (UK), and moves were initiated to lessen the role of nuclear forces in national security policy for some, all nine nuclear-armed states are doubling down on nuclear weapons as a central component of deterring nuclear-armed adversaries and competitors in the Third Nuclear Age. Moreover, as the next chapter explains, others may also wish to join the exclusive nuclear club in the future for exactly the same reason.

However, this is not playing out in a uniform way. Instead, there are three components to the evolving deterrence landscape in the Third Nuclear Age. First, all nine nuclear-armed states continue to see both political and military value in the threat of *retaliatory* nuclear strikes—all are therefore prioritising the surety of their secure second-strike capabilities ostensibly for deterrence through punishment. Most are replacing systems effectively like for like, while some are expanding nuclear options. This at least suggests a continuing belief in the theory of the nuclear revolution, albeit that what constitutes credibility may have shifted. Second, the Third Nuclear Age is likely to see a much greater test of the norm of nuclear non-use and the so-called nuclear taboo. A key part of this is the return of nuclear coercion, compellance, and threat making to international politics and a concomitant revival of thinking about the limited use of nuclear weapons on the battlefield. These aren't necessarily new dynamics—both were central to the First Nuclear Age—but today there are more players in more complex nuclear relationships than in the past. The third component, and perhaps the most significant, is the return to thinking about winning wars against

[3] Vipin Narang & Scott Sagan (eds.), *The fragile balance of terror* (Ithaca, NY: Cornell University Press, 2022), p. 3.

nuclear-armed opponents. What makes this particularly noteworthy is how technological change may be opening up the possibility in the eyes of some of deterring nuclear threats with advanced conventional weapons or 'Strategic Non-Nuclear Weapons'.[4] While deterring nuclear threats with advanced and networked non-nuclear platforms was something entertained by the US in the First Nuclear Age and pursued in the Second Nuclear Age, today technological advances and diffusion means there are more capable players than before. It has also reopened the question of conducing 'counterforce' strikes against nuclear targets—only this time without resorting to the use of nuclear forces. Of course, these pathways are not independent, and some states may be pursuing all three options simultaneously, but the typology does provide a framework for thinking about what is changing in the nuclear world around us.

The net result is that the ways, ends, and means of nuclear deterrence—and the pathways to nuclear security and strategic stability—are becoming more complex, blurred, and perhaps nuanced in the Third Nuclear Age. While different types of nuclear and conventional weapons systems and mission sets have always been somewhat entangled in deterrence strategies, the technological developments discussed in the previous chapter, and the geopolitical ones discussed in the next, have provided a tantalising glimpse of a new pathway to 'deterrence', but at the same time new challenges to proliferation and possibilities for coercion and escalation.

The enduring nuclear revolution?

Each of the nine current nuclear-armed states are undergoing programmes of modernisation and replacement of their nuclear forces. For some, this comes after decades where nuclear modernisation schedules were delayed and some key support infrastructure was allowed to slowly atrophy, for others nuclear modernisation has been a more gradual process over many years, and for a couple of states it has accelerated as we entered the Third Nuclear Age. This doesn't necessarily mean that significantly more nuclear weapons will be deployed, or that we will see arms racing akin to the First Nuclear Age, but it does suggest that qualitative and, in some cases, limited quantitative 'enhancements' to nuclear capabilities will be a prominent feature of

[4] Andrew Futter & Benjamin Zala, 'Strategic non-nuclear weapons and the onset of a Third Nuclear Age', *European Journal of International Security* 6:3 (2021), pp. 257–77.

global nuclear politics going forward, and a notable change from the Second Nuclear Age. The key reason for this appears to be attempting to make nuclear forces secure and ensuring that they can be credibly used if needed. This means keeping ahead of advances in missile and air defence systems, undersea warfare, precision strike counterforce capabilities, and possibly non-kinetic interference in nuclear command and control. Essentially, this logic suggests that nuclear modernisation is being driven by a more challenging technological deterrence environment, and the requirement to be able to hold adversary targets vulnerable.

The fact that the nuclear-armed states continue to modernise platforms to ensure the ability to threaten retaliation with nuclear weapons implies that classic ideas of stability based on vulnerability continue to hold. Essentially, it suggests that the threat of nuclear retaliation deters the use of nuclear weapons by an adversary because such action would be suicidal. This is central to the theory of the nuclear revolution: that nuclear weapons ultimately stabilise interstate relations due to mutual vulnerability to destruction.[5] Nevertheless, the criticisms of this model have not evaporated, and research continues to point to past near misses, the fallacy of relying on 'rational actor' models, and the role of luck. While impossible to quantify, the longer we go without a major nuclear weapons accident, or a freak event that leads to nuclear war, arguably the more likely it becomes. Nuclear deterrence is not foolproof.

The US is currently undergoing a multidecade, multibillion-dollar 'recapitalisation' of its nuclear weapons infrastructure, which became formalised under President Barack Obama in 2010. This includes building a new fleet of 400 Sentinel intercontinental ballistic missiles (ICBMs), a new nuclear-capable B21 stealth bomber, a long-range air-launched stand-off missile, the next generation of Columbia-class nuclear-powered and nuclear-armed submarines, as well as new nuclear warheads.[6] These form part of what is known as the 'program of record'.[7] According to the Congressional Budget Office in 2025, 'DoD's and DOE's plans to operate, sustain, and modernize current nuclear forces and purchase new forces would cost a total of $946 billion over the 2025–2034 period.'[8] (For comparison, that is just a

[5] E.g. Robert Jervis, *The meaning of the nuclear revolution* (Ithaca, NY: Cornell University Press, 1989).

[6] See '2022 nuclear posture review', https://media.defense.gov/2022/Oct/27/2003103845/-1/-1/1/2022-NATIONAL-DEFENSE-STRATEGY-NPR-MDR.pdf.

[7] This means that the programme and future funding have been officially approved.

[8] 'Projected costs of US nuclear forces, 2025–2034', Congressional Budget Office, April 2025, https://www.cbo.gov/publication/61362.

bit less than the size of Saudi Arabia's entire gross domestic product (GDP) in 2025).[9] Essentially, the US is replacing delivery platforms with updated versions rather than fundamentally changing what they are intended to do but taking into account new technological challenges to nuclear surety. This provides the US with a strategic triad, designed to guard against failure or compromise of any one nuclear 'leg'. As is discussed in the next chapter, this is designed to provide a credible strategic nuclear deterrent in a world of two nuclear-armed peer competitors: Russia and China, armed with advanced nuclear and non-nuclear weapons, and increasingly against North Korea and Iran. The US nuclear deterrent is also 'extended' to members of North Atlantic Treaty Organization (NATO), as well as to South Korea, Japan, and Australia—this is often cited as a reason why the US doesn't have a no-first-use or sole-purpose doctrine.[10]

Russia is coming to the end of a multidecade programme to replace Soviet era nuclear systems that began at the turn of the century, and has deployed a range of enhanced platforms: including the RS-28 SARMAT and the road-mobile RS-24 YARS ICBMs, the Avangard hypersonic glide vehicle, the Burevestnik nuclear-powered cruise missile, and the Poseidon nuclear-powered, nuclear-armed torpedo.[11] Russia has also modernised its nuclear-submarine fleet, and has begun deploying the 'Borei class' SSBN. These capabilities fall under the broader framework outlined in 'Fundamentals of State Policy of the Russian Federation of Nuclear Deterrence'.[12] A 2018 report by Julian Cooper for the Stockholm International Peace Research Institute (SIPRI) estimated that Russian nuclear weapons spending amounted to between $8 billion and £10 billion dollars per year, but this could have altered since the start of the Ukraine war in 2022.[13] At least some of these programmes have been justified by concerns about US missile defence and prompt strike ambitions, and the fear that the nuclear balance based on mutual vulnerability could be undermined. As President Putin put

[9] Saudi Arabia's GDP is estimated at $1.08 billion in 2025 according to the International Monetary Fund (IMF).

[10] Ankit Panda & Vipin Narang, 'Sole purpose is not no first use: Nuclear weapons and declaratory policy', *War on the Rocks*, 22 February 2021, https://warontherocks.com/2021/02/sole-purpose-is-not-no-first-use-nuclear-weapons-and-declaratory-policy/.

[11] Matthew Kroenig, Mark Massa, & Christian Trotti, 'Russia's exotic nuclear weapons and implications for the United States and NATO', *Atlantic Council Issue Brief*, March 2020, https://www.atlanticcouncil.org/wp-content/uploads/2020/07/Russias-Exotic-Nuclear-Weapons.pdf.

[12] 'Fundamentals of state policy of the Russia Federation on nuclear deterrence', 3 December 2024, https://www.mid.ru/print/?id=1434131&lang=en.

[13] Julian Cooper, 'How much does Russia spend on nuclear weapons?', *SIPRI*, 1 October 2018, https://www.sipri.org/commentary/topical-backgrounder/2018/how-much-does-russia-spend-nuclear-weapons.

it in his annual state of the nation speech in 2018, 'Russia has developed, and works continuously to perfect, highly effective but modestly priced systems to overcome missile defence'.[14] As of 2024, the team at the Federation of American Scientists estimates that Russia still has around 4,300 nuclear warheads in the stockpile—with approximately 2,800 allocated for 'strategic' missions, although since February 2026 deployed systems are no longer limited to 1,550 under the now defunct New Strategic Arms Reduction Treaty (START).[15]

The UK is in the process of replacing its fleet of SSBNs with the newer Dreadnought class—these were originally scheduled to begin being deployed in the early 2020s, but will probably now be deployed in the mid-2030s. The UK operates a policy of continuous-at-sea-deterrence whereby one of these submarines is always at sea ready to fire its missiles. The UK nuclear programme is closely linked with the US': engineers and scientists at the Atomic Weapons Establishment near Aldermaston in Berkshire are in the process of designing a new nuclear warhead in conjunction with the US, and the UK continues to rely on US-built Trident D5 submarine-launched ballistic missiles (SLBMs) for nuclear delivery. In 2021, the UK government then led by Boris Johnson announced that it would raise the maximum number of nuclear weapons held in the stockpile to 'no more than 260' and end the policy of transparency about the number of nuclear weapons deployed.[16] These decisions were probably the result of concerns about advances in Russian ballistic missile defence (BMD) and anti-submarine warfare (ASW) rather than any desire to enhance the role of nuclear forces in UK security, but nevertheless marked a notable break from several decades of nuclear reductions.[17] As Prime Minister Keir Starmer declared shortly after taking office in April 2024, 'In the face of rising global threats and growing Russian aggression, the UK's nuclear deterrent is the bedrock of Labour's plan to keep Britain safe.'[18] Along with the US, the UK commits its nuclear force

[14] Vladimir Putin, Presidential address to the Federal Assembly, 1 March 20218, http://www.en.kremlin.ru/events/president/transcripts/messages/56957.

[15] Hans M. Kristensen, Matt Korda, Eliana Johns, & Mackenzie Knight, 'Russian nuclear weapons, 2024', *Bulletin of the Atomic Scientists* 80:2 (2024), pp. 118–45.

[16] See 'Global Britain in a competitive age: The integrated review of security, defence, development and foreign policy', CP 403, UK Government, March 2021, https://assets.publishing.service.gov.uk/media/60644e4bd3bf7f0c91eababd/Global_Britain_in_a_Competitive_Age-_the_Integrated_Review_of_Security__Defence__Development_and_Foreign_Policy.pdf.

[17] 'UK nuclear warhead increase prompted by Russia's missile defence capability', *Financial Times*, 21 March 2021, https://www.ft.com/content/a86e8ca8-365e-4774-b22c-fbdf12237935.

[18] Quoted in Kate Whannel & Chas Geiger, 'Keir Starmer: Labour commitment to nuclear weapons unshakable', *BBC*, 2 April 2024, https://www.bbc.co.uk/news/uk-politics-68790435.

to the defence of NATO. It is difficult to get an exact figure on the cost of the UK nuclear weapons programme, but a reasonable estimate is that building the new submarines will cost between £31 and 41 billion, and in 2025 £15 billion was allocated to the replacement warhead programme.[19]

France is modernising both its sea-based and air-based nuclear forces—estimates suggest that this will cost around €6 billion (~$7 billion) per year at 2025 levels.[20] This includes the M51.3 and M51.4 SLBMs, four new SSBNs, a new ASN4G hypersonic air-launched missile and a new fleet of Rafale F5 jets.[21] Unlike the UK, the French nuclear force is entirely independent. It is being modernised for greater stealth, survivability, and the ability to penetrate adversary missile defences. France has traditionally been the most transparent nuclear-armed state, and has a total stockpile of approximately 290 nuclear warheads as of 2025, the majority of which are for use on submarine-launched ballistic missiles, and about 50 for delivery by aircraft.[22] Like the UK, the French government is keen to stress that this is the *minimum* capability necessarily.[23] As President Emmanuel Macron explained in a 2020 speech:

> France has a unique track record in the world, in keeping with its responsibilities and interests, having dismantled irreversibly its land-based nuclear component, its nuclear test facilities, its fissile material for weapons production facilities, and having reduced the size of its arsenal, which is currently under 300 nuclear weapons. These decisions are in line with our rejecting any type of arms race and our keeping the format for our nuclear deterrent at a level of strict sufficiency.[24]

[19] Clare Mills, 'The cost of the UK's strategic nuclear deterrent', House of Commons Library, Research Briefing, 12 August 2025, https://researchbriefings.files.parliament.uk/documents/CBP-8166/CBP-8166.pdf.

[20] 'Hidden costs: Nuclear weapons spending in 2024', ICAN report, June 2025, p. 6, https://assets.nationbuilder.com/ican/pages/8192/attachments/original/1749755658/ICAN_Spending_Report_Hidden_Costs_final_spreads.pdf?1749755658.

[21] Hans M. Kristensen, Matt Korda, & Eliana Johns, 'French nuclear weapons, 2023', *Bulletin of the Atomic Scientists* 79:4 (2023), pp. 272–81.

[22] Kristensen et al., 'French nuclear weapons, 2023'.

[23] As paragraph 168 on the National Security Review puts it: 'Nuclear deterrence is the ultimate guarantee of the Nation's security, protection and independence. France underlines the role and political nature of this weapon, whose use is the sole responsibility of the President of the Republic, elected by direct universal suffrage.' 'Revue nationale stratégique 2025', Secrétariat Général de la Défense et de la Sécurité Nationale (SGDSN), 2025, https://www.sgdsn.gouv.fr/files/files/Publications/20250713_NP_SGDSN_RNS2025_EN_0.pdf.

[24] Speech of the president of the Republic on the Defense and Deterrence Strategy, 7 February 2020, https://www.elysee.fr/en/emmanuel-macron/2020/02/07/speech-of-the-president-of-the-republic-on-the-defense-and-deterrence-strategy.

While French nuclear weapons are not part of the NATO nuclear deterrence mission, President Macron has begun to stress the role of French nuclear forces in European deterrence (discussed in more detail in the next chapter).

While the French and British nuclear capability has changed little since the Second Nuclear Age (with both initially reducing warheads and retiring delivery systems), the Chinese strategic nuclear modernisation programme has been and is comprehensive. The most significant developments have been in delivery systems: this includes the road-mobile DF-41 and DF-31 ICBMs, the JL3 SLBM, a new fleet of Type 94 and later Type 96 Jin class SSBNs, the H20 stealth bomber, and the DF-ZF hypersonic glide vehicle. But it may also include a massive expansion of the strategic stockpile of nuclear warheads (from about ~300 today to maybe closer to 1,000 by the end of this decade).[25] Beginning in 2019, it was reported that the Peoples Liberation Army Rocket Force (or PLARF) was constructing over 300 new ICBM silos across remote northern and western regions of the country.[26] This led some to suggest that the Chinese leadership was moving away from a 'minimum' nuclear force structure of the past towards a diverse and survivable force more similar to the US' and Russia's.[27] Part of the reason for this may be to reach parity in the number of deployed nuclear weapons with Washington and Moscow, but it is also likely to be driven by perceived requirements of nuclear survivability and thinking about how to deter US intervention in future regional conflicts. As leading expert Tong Zhao explains, this nuclear build-up is partly driven by concerns about US missile defence and precision strikes, but also about demonstrating strength.[28] Estimates suggest that China is spending about $10 billion per year on its nuclear weapons programme.[29]

India is modernising and slowly enhancing its nuclear force to meet the challenges posed by two nuclear-armed foes: Pakistan to the west and China to the east. Both present slightly different deterrence challenges.

[25] E.g. Paul Mcleary, 'Pentagon report: China to build 1000 nukes by 2030', *Politico*, 18 December 2024, https://www.politico.com/news/2024/12/18/pentagon-report-china-nuclear-weapons-00195031.

[26] Hans Kristensen & Matt Korda, 'China's nuclear missile silo expansion: From minimum deterrence to medium deterrence', *Bulletin of the Atomic Scientists*, 1 September 2021, https://thebulletin.org/2021/09/chinas-nuclear-missile-silo-expansion-from-minimum-deterrence-to-medium-deterrence/.

[27] E.g. Andrew F. Krepinevich Jr, 'The new nuclear age: How China's growing nuclear arsenal threatens deterrence', *Foreign Affairs* 101:3 (2022), pp. 92–104.

[28] Tong Zhou, 'What is driving China's nuclear build-up?', Carnegie Endowment for International Peace, 5 August 2021, https://carnegieendowment.org/posts/2021/08/whats-driving-chinas-nuclear-buildup?lang=en.

[29] 'Hidden costs', ICAN report, p. 6.

This is perhaps why India is developing a broad range of delivery systems. According to Hans Kristensen and the team at the *Nuclear Notebook*:

> We estimate that India currently operates eight different nuclear-capable systems: two aircraft, five land-based ballistic missiles, and one sea-based ballistic missile. At least five more systems are in development, most of which are thought to be nearing completion and to be fielded with the armed forces soon.[30]

This includes the deployment of the AGNI-V and VI ICBMs, K5 and K6 SLBMs, and the Arihant and S5 SSBNs. Indian authorities have stressed that the Brahmos hypersonic missile that can be delivered by land, air, and sea is not nuclear capable—but some experts disagree.[31] Together these platforms will give India a credible nuclear triad of delivery systems like those operated by the US, Russia, and China. It is estimated that as of 2025, India possesses around 170 warheads, but it is possible that the Indian nuclear stockpile will continue to grow over the next decade.[32] This suggests a move towards a more comprehensive nuclear deterrence strategy with global reach after decades wedded to a policy of 'minimum credible deterrence'. Estimating what India spends on its nuclear force is difficult,[33] but International Campaign to Abolish Nuclear Weapons (ICAN) estimated in 2024 that New Delhi spent around $2.6 billion on the nuclear complex, about 2.3% of the defence budget.[34] That said, there does not appear to be any move to match purported increases in Chinese nuclear capabilities.

Trying to distinguish platforms and mission sets in the Pakistani nuclear posture is a bit more difficult, partly because, as former director-general (DG) of the Strategic Plans Division (SPD), Lt General Kalid Kidwai explained in 2023, Pakistan's full-spectrum deterrent consists of 'strategic, operational and tactical' effects.[35] It is also complicated because of

[30] Hans M. Kristensen, Matt Korda, Eliana Johns, & Mackenzie Knight, 'Indian nuclear weapons, 2024', *Bulletin of the Atomic Scientists* 80:5 (2024), p. 326.

[31] E.g. Usman Haider, 'How the extended range Brahmos changes the India-Pakistan deterrence equation', *The Diplomat*, 7 November 2023, https://thediplomat.com/2023/11/how-the-extended-range-brahmos-changes-the-india-pakistan-deterrence-equation/.

[32] Kristensen et al., 'Indian nuclear weapons, 2024'.

[33] Urvashi Sarkar, 'What's known—and not known—about India's nuclear weapons budget', *Bulletin of the Atomic Scientists*, 2 November 2021, https://thebulletin.org/2021/11/whats-known-and-not-known-about-indias-nuclear-weapons-budget/.

[34] 'Hidden costs', ICAN report, p. 7.

[35] Speech by Lt Gen. (Retd) Khalid Kidwai, Advisor, National Command Authority and former DG SPD, on 25th Youme-e-Takbeer, 26 May 2023, https://issi.org.pk/speech-by-lt-gen-retd-khalid-kidwai-advisor-national-command-authority-and-former-dg-spd-on-25th-youme-e-takbeer/.

geography—Pakistan shares a land border with its main rival, and Islamabad and Delhi are only 400 miles apart, which means platforms could be used against different types of targets. But it is likely that the development of the road-mobile Shaheen ballistic and Babur cruise missiles as well as a Barbur-3 sea-launched cruise missile (SLCM) and Ra'ad air-launched cruise missile (ALCM), and the purchase of Chinese JF17 aircraft are designed for retaliatory strikes against 'countervalue' targets.[36] It is also thinkable that to meet growing requirements, Pakistan could possibly expand its nuclear warhead stockpile from ~170–180 as of 2025 to more than 200 over the next decade, although this is debated.[37] Part of the pressure driving expansion and modernisation of strategic nuclear forces are concerns about Indian missile defence deployments that might be used to protect strategic targets (major cities or military installations).[38] ICAN has estimated that Pakistan spends about $1 billion a year on its nuclear programme—about 0.3% of gross domestic product (GDP).[39]

Deciphering the mission sets and budget for Israeli nuclear forces is also difficult due to the secrecy that surrounds the programme, but evidence suggests that the stockpile of around 90 warheads and nuclear delivery platforms is for retaliation—often referred to as the 'Samson option' of last resort.[40] Although at the time of writing, there aren't any other nuclear-armed states in the Middle East. Israel has missiles and aircraft that can reach most parts of the region, notably Teheran, but also longer range missiles with global reach.[41] This includes the Jericho II and III land-based ballistic missiles, nuclear-capable F15, F16, and possibly F35 aircraft, and maybe the Popeye SLCM onboard Dolphin-class air-independent propulsion submarines.

The ability to strike the US homeland with nuclear weapons has always been a central component of the North Korean missile programme. For

[36] Hans M. Kristensen, Matt Korda, & Eliana Johns, 'Pakistan nuclear weapons, 2023', *Bulletin of the Atomic Scientists* 79:5 (2023), pp. 329–45. Countervalue being the polite way of saying 'population centres'.

[37] Kristensen et al., 'Pakistan nuclear weapons, 2023'.

[38] On this, see Zafar Khan, 'India's ballistic missile defense: Implications for south Asian deterrence stability', *The Washington Quarterly* 40:3 (2017), pp. 187–202.

[39] 'Hidden costs', ICAN report, p. 6.

[40] For a historical background to this, see Seymour M. Hersch, *The Samson option* (Random house, New York: 1991); and Avner Cohen, *Israel and the bomb* (Columbia University Press: 1998).

[41] Getting contemporary data on Israeli nuclear forces is very difficult. By far the best source is: Hans M. Kristensen & Matt Korda, 'Israeli nuclear weapons, 2021', *Bulletin of the Atomic Scientists* 78:1 (2022), pp. 38–50.

most of the Second Nuclear Age, Pyongyang probably wasn't able to do this, but beginning in the late 2010s a range of ICBMs were deployed that could constitute a credible second strike. These include the Hwasong 15, 17, and 18. Even with various US and allied BMD systems deployed and possible pre-emptive options being pursued, this has changed the deterrence game for the military and political leadership around Kim Jong-un. While North Korea is pursuing a variety of different missile delivery platforms for different purposes, an ability to deter the US by threatening the homeland has remained paramount.[42] While very difficult to assess, estimates suggest that North Korea spends about £630 million on nuclear weapons (which would be around 2.5% of GDP).[43]

This all implies that a central feature of the Third Nuclear Age will be a doubling down on the belief in nuclear weapons as a central way to deter nuclear attack. Most states are doing this to enhance secure second-strike nuclear capabilities—that is the ability to threaten punishment in response to any nuclear attack. This in turn is being shaped by developments in a range of systems that might undermine the perceived viability of these platforms such as better air and missile defences, the spread of precision missiles, and enhanced underwater sensing and ASW. Thus, the belief that stability can be enhanced through mutual vulnerability at the nuclear level remains pervasive—at least amongst the 'nuclear elites'. It is possible that some $100 billion was spent on nuclear weapons programmes in 2025 alone—about the same amount that Bill Gates has given away to charitable causes in the past 25 years.[44]

Battlefield nuclear weapons, limited war, and the end of the taboo?

In addition to the technological and geopolitical pressures that are cited as driving modernisation of 'secure second-strike' nuclear forces, a concurrent trend is also emerging in the Third Nuclear Age: the return of thinking about nuclear weapons in war. This refers to a suite of capabilities that might be used against military targets in military situations rather than against large

[42] Hans M. Kristensen, Matt Korda, Eliana Johns, & Mackenzie Knight, 'North Korean nuclear weapons 2024', *Bulletin of the Atomic Scientists* 80:4 (2024), pp. 251–71.

[43] 'Hidden costs', ICAN report, p. 6.

[44] Bill Gates, 'My new deadline: 20 years to give away virtually all my wealth', *Gatesnotes*, 8 May 2025, https://www.gatesnotes.com/n20-years-to-give-away-virtually-all-my-wealth.

population centres or against other *strategic* targets (although there is no one agreed definition of this). Of course, so-called tactical or substrategic nuclear forces were central to First Nuclear Age deterrence thinking, and especially to concepts of how the US and its allies might to deter war in Europe or Asia,[45] and—when combined with Soviet tactical weapons—may have numbered as many as 20,000–30,000 in the late 1980s.[46] But a purported belief in the coercive/compellent power of limited nuclear threats and perhaps a concurrent belief in the utility of 'smaller' nuclear explosions as a means of deterring or fighting a conventional war have returned.[47] Importantly, it is not just the traditional 'baddies', as the post-Cold War Western construct might have it, that appear interested in conceptions of limited nuclear war—or at least 'escalation management options', but also the US and NATO.

The belief in the utility of nuclear use on the battlefield was a central component of the Cold War, and of both NATO and Soviet strategic thinking. In the early Cold War, short-range nuclear weapons were seen as compensation for Western conventional weakness in Europe, and as a way of cost cutting, but also as a means of fighting a war should it break out. In theory any nuclear weapon platform could be used 'tactically'—the term refers more to the target and warhead yield rather than any particular system; but the tactical nuclear weapon (TNW) label tended to refer to munitions such as land mines, short-range ballistic missiles, depth charges, torpedo's, bombs dropped by tactical aircraft, artillery shells, cruise missiles, and even the 'Davy Crockett' short-range recoilless launcher system (which had a range of just 1–2 miles!). Hundreds of US, Soviet (and a few British) TNWs were deployed on ships at sea, with troops, and at air bases across Europe and Asia (sometimes without the host country being aware).[48]

The deployment of battlefield nuclear weapons raised a number of different concerns to those associated with 'strategic' forces. First, in some cases TNWs were deployed outside the country of ownership, and were designed to be used quickly, and probably before the opponent had used nuclear weapons. The classic example being use by NATO against a conventional

[45] E.g. Catherine McArdle Kelleher, 'Managing NATO's tactical nuclear operations', *Survival* 30:1 (1988), pp. 59–78.

[46] Hans M. Kristensen & Matt Korda, 'Tactical nuclear weapons, 2019', *Bulletin of the Atomic Scientists* 75:5 (2019), pp. 252–61.

[47] Deterrence is best thought of as seeking to *prevent* an action; compellance/coercion is seeking to *force* a particular action.

[48] See Robert S. Norris, William M. Arkin, & William Burr, 'Where they were', *Bulletin of the Atomic Scientists* 55:6 (1999), pp. 26–35.

invasion of western Europe. In some cases, this meant pre-delegating the authority to use these weapons to commanders in the field to ensure deterrence credibility.[49] Second, the use of limited tactical nuclear operations raised the question of whether what would happen next could be 'controlled'. Would nuclear escalation follow an increasingly intolerable and disastrous upward trajectory towards bigger weapons and more targets, or could war be terminated before nuclear use spread? And third, could the threat of using 'smaller' battlefield nuclear weapons be used coercively—seeking to force (or in the jargon—compelling) an adversary into taking a particular course of action? History suggests that coercing or compelling with nuclear threats has been relatively ineffective, but this hasn't stopped various leaders from trying.[50]

With the end of the Cold War, the role and rationale for tactical nuclear forces in Western/NATO nuclear strategy began to fall away. In 1991, the US repatriated land-based nuclear arms from Europe and South Korea; in 1993 US tactical weapons were removed from surface ships and submarines, and in 1998 the UK retired its tactical WE177 free-fall bomb. At the same time, the continued basing of US nuclear gravity bombs deployed in Europe under the NATO nuclear sharing arrangement came under increased pressure. In 2009, the German government began a push to de-nuclearise the US TNW deployed in NATO Europe,[51] and in the early 2010s, there were even suggestions that the US and Russia might sign a treaty that reduced or even eliminated deployed tactical nuclear forces entirely. However, over the past decade or so, battlefield nuclear weapons have returned as a component of statecraft. Only this time there are more players involved and perhaps less restraint.

While ending the continuous deployment of several types of land- and sea-based nuclear weapons in the early 1990s, Russia continues to hold thousands of tactical nuclear warheads in its inventory and has updated and upgraded multiple possible delivery platforms.[52] These include short-range

[49] See William Burr, 'First declassification of Eisenhower's instructions to commanders pre-delegating nuclear weapons use, 1959–1960', *The National Security Archive*, 19 May 2001, https://nsarchive2.gwu.edu/NSAEBB/NSAEBB45/printindex.html.

[50] See Todd S. Sechser & Matthew Fuhrmann, *Nuclear weapons and coercive diplomacy* (Cambridge: Cambridge University Press, 2017).

[51] Oliver Meier, 'German stance stirs nuclear debate', *Arms Control Today*, December 2009, https://www.armscontrol.org/act/2009-12/german-nuclear-stance-stirs-debate.

[52] These have a multitude of functions from air and missile defence to anti-ship and anti-submarine warfare and coastal defence as well as bombs and missiles for land attack. See Hans M. Kristensen, Matt Korda, Eliana Johns, & Mackenzie Knight, 'Russian nuclear weapons, 2025', *Bulletin of the Atomic Scientists* 81:3 (2025), pp. 208–37.

missiles and air-launched missiles such as the Iskander-M and Kinzhal, as well as nuclear bombers and artillery shells. Many of these platforms are also dual use—meaning that they might be 'mated' with either a conventional or nuclear warhead. Since 2024, Russia is also believed to deploy tactical nuclear platforms in Belarus.[53] While modernisation of the Russian tactical nuclear capability has been ongoing for a generation, the overt return to using nuclear threats is something that re-emerged alarmingly in 2022. This reached a peak in October of that year when reports surfaced that Russia may be considering using nuclear weapons against Ukrainian forces. As President Putin said at the time, 'In the event of a threat to the territorial integrity of our country and to defend Russia and our people, we will certainly make use of all weapon systems available to us. This is not a bluff.'[54] Periodically, concerns have also emerged that Russia might seek to carry out a limited nuclear strike against military targets, including the Ukrainian leadership.[55] These fears have been bolstered by a perception that Russia may be willing to 'escalate to de-escalate'—in effect using a limited nuclear strike in a conventional conflict to force adversaries (such as NATO or the US) to back down rather than escalate further.[56]

At least in part due to the nuclear tensions that have accompanied the Ukraine war, thinking about battlefield nuclear applications has also returned to NATO strategic planning. The US B61 nuclear gravity bomb has been upgraded, and Belgium, Germany, the Netherlands, and Italy have all agreed to purchase the US-built dual capable F35a aircraft to deliver them.[57] In 2025, the Strategic Defence Review recommended that the UK should buy some F-35a aircraft and commence discussions with the US and NATO about joining NATO's Dual Capable Aircraft (DCA) mission.[58]

[53] Liviu Horovitz & Lydia Wachs, 'Russian nuclear weapons in Belarus? Motivations and consequences', *The Washington Quarterly* 47:3 (2024), pp. 103–29.

[54] 'Address by the president of the Russian Federation', Kremlin, 21 September 2022, http://en.kremlin.ru/events/president/transcripts/69390.

[55] David E. Sanger, 'Biden's Armageddon moment: When nuclear detonation seemed possible in Ukraine', *The New York Times*, 9 March 2024, https://www.nytimes.com/2024/03/09/us/politics/biden-nuclear-russia-ukraine.html.

[56] Nikolai Sokov, 'Russian military doctrine calls a limited nuclear strike "de-escalation". Here's why', *Bulletin of the Atomic Scientists*, 8 March 2022, https://thebulletin.org/2022/03/russian-military-doctrine-calls-a-limited-nuclear-strike-de-escalation-heres-why/.

[57] The ability for Turkey to participate in this mission is uncertain as the Turkish government is (at the time of writing) not part of the F35 programme, though it is believed that US B61 bombs are still based at Incirlik air base in the south of the country. See Hans Kristensen, Matt Korda, Eliana Johns, & MacKenzie Knight, 'Nuclear weapons sharing, 2023', *Bulletin of the Atomic Scientists* 79:6 (2023), pp. 393–406.

[58] 'Making Britain safer: Secure at home, strong abroad', Strategic Defence Review, Ministry of Defence, 2025, https://assets.publishing.service.gov.uk/media/683d89f181deb72cce2

Three weeks later Sir Keir Starmer's government announced that it would do exactly that.[59] This followed a short period where debate had returned about whether the UK needed to re-establish an additional tactical nuclear platform.[60] In the past, UK nuclear-armed submarine-launched ballistic missiles were described as having a 'substrategic' role and could be targeted against military forces—although this would present possible problems given the reliance on one platform.[61] France deploys an air-delivered weapons system that could be used in a tactical/non-strategic way, but French officials are very clear that they reject the notion of battlefield applications.[62]

While tactical nuclear platforms were de-emphasised for a generation in the US, there seems to be increasing acceptance that battlefield applications might be needed to enhance deterrence and for fighting a war (either in Europe or Asia). In 2020, the US began deploying the 5 kiloton 'low-yield Trident' W76-2 warhead, while renewed calls for a SLCM-N as well as continued investment in the B61 warhead and stealthy F35a aircraft reflect this.[63] The justification of a new sea-based platform is probably linked to perceived deterrence requirements against China. It is also possible that US TNWs might return (they were removed in 2008) to the US air base at Royal Air Force (RAF) Lakenheath in the UK—possibly linked with the decision by the UK government to buy F35a aircraft and possibly rejoin the DCA mission ...[64] As the respected experts Jeffrey Larsen and Kerry Kartchner warned over a decade ago: 'even if the United States wants to avoid a limited

680a5/The_Strategic_Defence_Review_2025_-_Making_Britain_Safer_-_secure_at_home__strong_abroad.pdf.

The UK had having previously purchased only the non-nuclear F35b variant.

[59] 'UK to purchase F-35As and join NATO nuclear mission as government steps up national security and delivers defence dividend', UK Government, https://www.gov.uk/government/news/uk-to-purchase-f-35as-and-join-nato-nuclear-mission-as-government-steps-up-national-security-and-delivers-defence-dividend.

[60] See 'How could the UK augment its nuclear forces', Council on Geostrategy, 28 March 2024, https://www.geostrategy.org.uk/britains-world/how-could-the-uk-augment-its-nuclear-forces/.

[61] See Malcolm Chalmers, '"Bombs away"? Britain and Nuclear weapons under New Labour', *Security Dialogue* 30:1 (1999), pp. 61–74.

[62] This was pointed out to me several times in off-the-record interviews with French officials.

[63] E.g. John Harvey & Robert Soofer, 'Strengthening deterrence with SLCM-N', *Atlantic Council*, 5 November 2022, https://www.atlanticcouncil.org/in-depth-research-reports/issue-brief/strengthening-deterrence-with-slcm-n/.

[64] In August 2023, it was rumoured that nuclear-capable US aircraft were returning to bases in the UK. Matt Korda & Hans Kristensen, 'Increasing evidence that the US air force's nuclear mission may be returning to UK soil', Federation of American Scientists, 28 August 2023, https://fas.org/publication/increasing-evidence-that-the-us-air-forces-nuclear-mission-may-be-returning-to-uk-soil/.

nuclear war, it may nevertheless find itself in one not of its choosing. Each side gets a vote on how the conflict develops.'[65]

In addition to the *return* of TNWs to US, NATO, and Russian thinking, the possibility of battlefield applications has now become part of Pakistani and North Korean strategy. There are slightly different reasons for this: the belief that nuclear forces can offset a conventionally superior opponent, that such forces could possibly be used for coercion, as well as more abstract ideas about escalation management. While the US, Russia, and NATO appear to be rejuvenating past ideas, it is possible that leaders in Pakistan and North Korea are drawing inspiration from the same Cold War era playbook.[66]

The best example is perhaps the Pakistani deployment of the road-mobile NASR short-range missile (with an estimated range of 40 miles), which has purportedly been designed to counter India's 'Cold Start doctrine' by targeting conventional military forces.[67] According to Pakistani scholar Mansoor Ahmed, it is also possible that Pakistan has developed nuclear artillery munitions similar to the W-33 nuclear artillery shell built by the US in the 1960s.[68] Slightly less widely reported are moves by North Korea to deploy tactical nuclear platforms that could target US, South Korean, or Japanese forces on land or at sea, including short-range missiles and rockets. Kim Jong Un began speaking about 'tactical nuclear weapons' in the early 2020s, but as the US expert Ankit Panda explains, it is not entirely clear what this means. From the outside, the logic would appear to be that such weapons would be used to deter a conventional attack, while attacks on soft targets such as population centres would be held back.[69] Given disparities in conventional capabilities, this likely makes the threshold for tactical nuclear use

[65] Jeffrey A. Larsen & Kerry M. Kartchner, *On limited nuclear war in the 21st century* (Stanford, CA: Stanford University Press, 2014), p. 4.

[66] Sadia Tasleem & Toby Dalton, 'Nuclear emulation: Pakistan's nuclear trajectory', *The Washington Quarterly* 41:4 (2018), pp. 135–55.

[67] Ben Barry, 'Pakistan's tactical nuclear weapons: Practical drawbacks and opportunity costs', *Survival* 60:1 (2018), pp. 75–81.

However, according to Barry: 'In war, the practical utility of Pakistan's short-range, low-yield nuclear capabilities may be more problematic than is often assumed.'

[68] See Mansoor Ahmed, 'Pakistan's tactical nuclear weapons and their impact on stability', Carnegie Endowment for International Peace, 30 June 2016, https://carnegieendowment.org/research/2016/06/pakistans-tactical-nuclear-weapons-and-their-impact-on-stability?lang=en.

[69] Ankit Panda, 'A call to arms: Kim Jong Un and the tactical bomb', *The Washington Quarterly* 44:3 (2021), pp. 7–24.

very low.[70] It is less clear what Chinese thinking is regarding TNWs.[71] Along with India, China has a formal no-first-use doctrine, and both state never to use nuclear weapons first—but technically both have missiles and bombs that could be used in a tactical/non-strategic capacity.[72]

Perhaps what makes this resurgence in tactical, battlefield, or non-strategic nuclear applications even more worrying is that they have been accompanied by a loosening of nuclear rhetoric and in some cases overt nuclear threat making. President Donald Trump's taunt to North Korean leader Kim Jong-un in 2018 about 'having a "bigger nuclear button" ... that is much bigger and more powerful'[73] and purported threats to carry out a pre-emptive nuclear attack against the regime in 2017;[74] Indian Prime Minister Narendra Modi's declaration a year later that 'Indian nuclear weapons were not being saved for Diwali',[75] or various North Korean propaganda videos that have been released of nuclear weapons destroying parts of the US.[76] But the major concern has been the nuclear dimension of the war in Ukraine, and particularly a mixture of thinly veiled and sometimes overt threat making by the Kremlin and across Russian media.

[70] Adam Mount, 'North Korea's tactical nuclear threshold is frighteningly low', *Foreign Policy*, 8 December 2022, https://foreignpolicy.com/2022/12/08/north-korea-tactical-nuclear-threat/.

[71] David Logan, 'The dangerous myths about China's nuclear weapons', *War on the Rocks*, 18 September 2020, https://warontherocks.com/2020/09/the-dangerous-myths-about-chinas-nuclear-weapons/.

[72] E.g. Kumar Sundaram & M.V. Ramana, 'India and the policy of no first use of nuclear weapons', *Journal for Peace and Nuclear Disarmament* 1:1 (2018), pp. 152–68.

[73] The tweet in full: 'North Korean Leader Kim Jong Un just stated that the "Nuclear Button is on his desk at all times". Will someone from his depleted and food starved regime please inform him that I too have a Nuclear Button, but it is a much bigger & more powerful one than his, and my Button works!'. See Lauren Gambino, 'Donald Trump boasts that his nuclear button is bigger than Kim Jong-un's', *The Guardian*, 3 January 2018, https://www.theguardian.com/us-news/2018/jan/03/donald-trump-boasts-nuclear-button-bigger-kim-jong-un.

[74] Rebecca Shabad, 'Trump discussed using a nuclear weapon on North Korea in 2017 and blaming it on someone else, book says', *NBC News*, 12 January 2023, https://www.nbcnews.com/politics/donald-trump/trump-discussed-using-nuclear-weapon-north-korea-2017-blaming-someone-rcna65120.

[75] 'India has stopped the policy of getting scared by Pakistan's threats. Every day, they (Pakistan) would make claims about having nuclear weapons, Even the media would bring out reports about Pakistan having nuclear weapons. So, what do we have? Are we saving them for Diwali'; Palak Nandi & Vimal Bhatia, 'Our nukes not for Diwali: PM Narendra Modi on Pakistan's N-threat', *Times of India*, 22 April 2019, https://timesofindia.indiatimes.com/elections/news/our-nukes-not-for-diwali-pm-on-pakistans-n-threat/articleshow/68982495.cms.

[76] E.g. 'North Korea nukes US in propaganda video', *Sky News*, https://news.sky.com/story/north-korea-nukes-us-in-propaganda-video-10218740.

In mid-2024, a video depicting a nuclear strike on the UK was aired on Russian TV. The video was presumably meant as a warning to the UK and others purportedly ramping up support to Ukraine.[77] Such posturing has become a regular feature of Russian media, with one expert recently commenting that 'Nuclear threats have become normalised in Russian civil society—everyone who's anyone has had something to say'.[78] Another respected expert has privately described the influence of a group of 'nuclear fire worshippers' surrounding President Putin who are keen to demonstrate Russian resolve.[79] Official reference to nuclear threats has been slightly more subtle, tending to refer to Russian nuclear doctrine or certain red lines. These numerous remarks have been catalogued by a team led by Heather Williams at the Center for Strategic and International Studies.[80]

Tactical platforms and the perception of a lessening of rhetorical restraint are beginning to impact two central pillars of 'nuclear order': the norm of non-use and the nuclear taboo. The norm of non-nuclear use is usually associated with the Canadian political scientist T.V. Paul and refers to the fact that nuclear use has rarely made political or military sense.[81] The taboo—a termed popularised by US academic Nina Tannenwald in 1999—is slightly different in that it refers to the moral and ethical drivers of non-nuclear use that have developed over time.[82] Together these ideas formed a strong belief, particularly in the Second Nuclear Age, that nuclear weapons could have no role in warfare. Indeed, over two decades later in 2022, the leaders of the 'P5' nuclear states agreed the following statement:

> We affirm that a nuclear war cannot be won and must never be fought. As nuclear use would have far-reaching consequences, we also affirm that

[77] 'Briefing: Russian state TV renews nuclear threats against West', *BBC Monitoring*, 29 April 2024, https://monitoring.bbc.co.uk/product/b0001j0s.

[78] Remarks at an event under Chatham House rule.

[79] Private conversation with Russian expert.

[80] Heather Williams, Kelsey Hartigan, Lachlan MacKenzie, & Reja Younis, 'Deter and divide: Russia's nuclear rhetoric & escalation risks in Ukraine', Centre for Strategic and International Studies, (11 January 2024), https://nuclearrussiaukraine.csis.org/#about.

[81] T.V. Paul, *The tradition of the non-use of nuclear weapons* (Stanford, CA: Stanford University Press, 2009).

As Paul put it, 'The tradition seems to have emerged due to the realizations that nuclear weapons are radically different from conventional weapons and that their use would show the user in an excessively bad light in international public opinion'. p. 3.

[82] Nina Tannenwald, 'The nuclear taboo: The United States and the normative basis of nuclear non-use', *International organization* 53:3 (1999), pp. 433–68. See also Nina Tannenwald, *The Nuclear Taboo: The United States and the Non-Use of Nuclear Weapons Since 1945* (Cambridge: Cambridge University Press, 2007).

nuclear weapons—for as long as they continue to exist—should serve defensive purposes, deter aggression, and prevent war.[83]

But even before Russia invaded Ukraine, Tannenwald had warned in 2018 that: 'Now more than ever before, humanity risks facing a future in which the nuclear taboo, a hard-won norm that makes the world a safer place, is in retreat'.[84] The fear is that any type of nuclear explosion, however 'limited' in its effects, may normalise nuclear use. As Tannenwald put it four years later, after the outbreak of the war, 'The nuclear taboo, while widely shared, is more fragile than other kinds of norms because a small number of violations would likely destroy it'.[85] That said, the fact that both India and China released strong statements condemning the possible use of nuclear weapons,[86] and that any NATO response would purportedly have been conventional,[87] suggests that the taboo and norm still have power.

Winning wars against nuclear-armed opponents?

While the planetary-level scale of destruction that nuclear war could bring about had gained considerable traction by the late 1950s, thanks in particular to Neville Schute's book (and later film) *On the Beach*, this did not prevent experts and officials from 'thinking the unthinkable' as US strategist

[83] 'Joint statement of the leaders of the five nuclear-weapon states on preventing nuclear war and avoiding arms races', 3 January 2022, https://bidenwhitehouse.archives.gov/briefing-room/statements-releases/2022/01/03/p5-statement-on-preventing-nuclear-war-and-avoiding-arms-races/#:~:text=We%20affirm%20that%20a%20nuclear,deter%20aggression%2C%20and%20prevent%20war.

[84] Nina Tannenwald, 'The vanishing nuclear taboo?', *Foreign Affairs* 97:6 (2018), pp. 16–24, at p. 18.

[85] Nina Tannenwald, 'Is using nuclear weapons still taboo?', *Foreign Policy*, 1 July 2022, https://foreignpolicy.com/2022/07/01/nuclear-war-taboo-arms-control-russia-ukraine-deterrence/.

[86] Stuart Lau, 'China's Xi warns Putin not to use nuclear arms in Ukraine', *Politico*, 4 November 2022, https://www.politico.eu/article/china-xi-jinping-warns-vladimir-putin-not-to-use-nuclear-arms-in-ukraine-olaf-scholz-germany-peace-talks/; and 'Indian defense minister warns parties to Ukrainian conflict against using nuclear weapons', *TASS*, 26 October 2022, https://tass.com/world/1757231.

[87] Jeffrey Edmonds, 'Potential US responses to the Russian use of non-strategic nuclear weapons in Ukraine', *Bulletin of the Atomic Scientists*, 16 May 2022, https://thebulletin.org/2022/05/potential-us-responses-to-the-russian-use-of-non-strategic-nuclear-weapons-in-ukraine/?utm_.

Herman Kahn put it his controversial 1960 book.[88] Such thinking was not necessarily driven by a desire to pre-emptively destroy an adversary in some dastardly plot, but rather was the result of attempts to contemplate and work through the mechanics of nuclear war and how it might be deterred, and, should war break out, piece together what 'victory' might look like. This led to the famous quote from Kahn that nuclear war might be 'winnable':

> Despite a widespread belief to the contrary, objective studies indicate that even though the amount of human tragedy would be greatly increased in the postwar world, the increase would not preclude normal and happy lives for the majority of survivors and their decedents.[89]

Kahn (perhaps crossed with the controversial physicist Edward Teller) was allegedly the basis for Stanley Kubrick's Dr Strangelove in the eponymous film released three years later.[90]

The result was that both the US and Russia developed doctrines and force structures not just for assured retaliation against 'soft' targets, but also options for counterforce strikes: that is the ability to destroy the adversaries' nuclear weapons and command systems before they could be used. Given the size of US and Russian nuclear arsenals throughout most of the Cold War, and the technological limitations of accuracy and real-time surveillance, this was a mammoth task. Nevertheless, ideas of 'countervailing' strategies captured by the acronym NUTs (short for National Utilisation Target Selection)[91] continued, and in 1979 Colin Gray called for a 'theory of victory' in a widely discussed article in the journal *International Security*.[92] These ideas were challenged in the mainstream,[93] and as former 'missileer' Bruce Blair later pointed out 'Exquisite nuclear warfighting is the stuff of

[88] Herman Kahn, *On thermonuclear war* (Piscataway, NJ: Transaction Publishers, 2007 [1960]).

[89] Kahn, *On thermonuclear war*, p. 21.

[90] Dorian Lynskey, 'Dr Strangelove at 60: The mystery behind Kubrick's Cold War masterpiece', *BBC Culture*, 29 January 2024, https://www.bbc.co.uk/culture/article/20240129-dr-strangelove-at-60-the-mystery-behind-kubricks-cold-war-masterpiece.

[91] See Spurgeon M. Keeny Jr & Wolfgang K.H. Panofsky, 'Mad versus nuts, *Foreign Affairs* 60:2 (Winter 1981–2), pp. 287–304.

'For convenience, and not in any spirit of trading epithets, we have chosen the acronym of NUTS to characterize the various doctrines that seek to utilize nuclear weapons against specific targets in a complex of nuclear war fighting situations intended to be limited ...' p.?

[92] Colin S. Gray, 'Nuclear strategy: The case for a theory of victory', *International Security* 4:1 (1979), pp. 54–87.

[93] Michael E. Howard, 'On fighting a nuclear war', *International Security* 5:4 (1981), pp. 3–17.

armchair strategists living in some parallel universe, not the real world'.[94] But the desire of escaping mutual assured destruction (MAD) never really went away in US and perhaps to a lesser extent Soviet thinking.[95] But this was not the same in the UK and France—where experts such as the former UK permanent undersecretary of state for defence, Sir Michael Quinlan, and the French strategic thinker Therese Delpech argued for deterrence focussed on minimal credible and ethical nuclear postures rather than warfighting.[96] Neither was it reflected in Maoist thinking about nuclear deterrence, which emphasised the political role of nuclear weapons, which in turn shaped Chinese doctrine throughout the First Nuclear Age.[97]

While after 1998 the move into a Second Nuclear Age ushered in the possibility of a regional nuclear war detached from the politics of the Cold War (between India and Pakistan), it also marked the beginning of a perception that the US might be on the verge of achieving 'nuclear primacy'. In 2006, American academics Keir Leiber and Daryl Press wrote that the US may be approaching 'the end of MAD'; as they put it: 'the United States now stands on the verge of attaining nuclear primacy, meaning that it could conceivably disarm the long-range nuclear arsenals of Russia and China with a nuclear first strike'.[98] A decade and a half later, Christopher Clary and Vipin Narang warned that India may be pursuing a possible nuclear counterforce capability against Pakistan.[99] But while the possibility of nuclear counterforce remains in the Third Nuclear Age, an arguably more pressing challenge relates to ever-improving *conventional* weapons.

While non-nuclear forces have always played a role in general deterrence, and to think of nuclear deterrence as somehow separate from the wider

[94] Bruce G. Blair, 'Loose cannons: The president and US nuclear posture', *Bulletin of the Atomic Scientists*, 1 January 2020, https://thebulletin.org/premium/2020-01/loose-cannons-the-president-and-us-nuclear-posture/.

[95] See Charles L. Glaser, 'The end of MAD? Technological innovation and the future of nuclear retaliatory capabilities', *Journal of Strategic Studies* 48:2 (2025), pp. 239–251.

[96] See, for example, Michael Quinlan, *Thinking about nuclear weapons: Principles, problems, prospects* (London: Royal United Services Institute for Defence and Security Studies RUSI, 2009); and Therese Delpeche, *Nuclear deterrence in the 21st century: Lessons from the Cold War for a new era of strategic piracy* (Santa Monica, CA: RAND Corporation, 2012).

[97] As Mao famously put it in an interview in 1946—nuclear weapons were 'paper tigers'. See Jeffrey Lewis, *Paper tigers: China's nuclear posture* (London: Routledge, 2014).

[98] Keir A. Lieber and Daryl Press, 'The end of MAD? The nuclear dimension of U.S. primacy', *International Security* 30:4 (Spring 2006), pp. 7–44.

[99] Christopher Clary & Vipin Narang, 'India's counterforce temptations: Strategic dilemmas, doctrine, and capabilities', *International Security* 43:3 (2019), pp. 7–52.

For an opposing view, see Rajesh Rajagopalan, 'India and counterforce: A question of evidence', *Observer Research Foundation* (14 May 2020), https://www.orfonline.org/research/india-and-counterforce-a-question-of-evidence.

strategic balance is misleading, a distinctly new pathway is emerging in the Third Nuclear Age. This is being driven by many of the technological advances discussed in the previous chapter and facilitated by the development of a suite of 'strategic non-nuclear weapons' or 'advanced conventional weapons' that are increasingly seen as being able to carry out mission sets previously reserved for nuclear forces.[100]

These mission sets include destroying hardened fixed strategic targets (such as ICBM silos, command bunkers, airfields, submarine pens, or radar's); tracking and holding vulnerable mobile nuclear platforms such as transport-erector-launcher missiles (TELs) and nuclear-armed submarines at sea; intercepting and shooting down warheads, missiles, and aircraft; and interfering in nuclear command, control, and communications (NC3). As discussed above, in the First Nuclear Age, all of these missions were largely reserved for nuclear forces—this was partly a function of accuracy, intelligence capabilities, and speed.[101] In the Second Nuclear Age this started to change as particularly the US began to pursue non-nuclear means of deterrence such as 'global prompt strikes' and layered missile defences to address 'rogue' nuclear challenges. In the Third Nuclear Age this phenomenon is becoming much more widespread, and, for some, may be seen as an alternative (rather than augmenting) pathway to address nuclear threats. This is a group that in a recent book the Third Nuclear Age team and I described as the 'technological transformers'.[102]

The incorporation of strategic non-nuclear platforms into nuclear deterrence strategy can probably be traced back to the 1970s. But it only really began to materialise after the Cold War had finished. Some scholars point to the 'revolution in military affairs' concept and the use of tactical precision weapons, space-based intelligence, surveillance, and reconnaissance (ISR), and the Patriot missile defence system in the 1991 Gulf War as the

[100] As Fabian Hoffman and William Alberque explain, 'Non-nuclear strategic weapons refer to weapons systems below the nuclear threshold that can achieve decisive strategic effects ... Put differently, non-nuclear strategic weapons allow their possessor states to bypass the tactical and operational levels of warfare, potentially enabling decisive victory prior to defeating the adversary's armed forces on the battlefield.' In Fabian Hoffman & William Alberque, '"Non-nuclear weapons with strategic effects" new tools of warfare?', *International Institute for Strategic Studies*, March 2022, p. 3, https://www.iiss.org/globalassets/media-library---content--migration/files/research-papers/2022/03/non-nuclear-weapons-with-strategic-effect-new-tools-of-warfare.pdf.

[101] Perhaps one exception being the use of submarines using sonar and conventionally armed torpedoes to find and sink nuclear-armed submarines. A point made to me recently by David Blagden.

[102] Andrew Futter, Ludovica Castelli, Cameron Hunter, Olamide Samuel, Francesca Silvestri, & Benjamin Zala, *The global Third Nuclear Age* (London: Routledge, 2025).

forerunners of today's strategic non-nuclear platforms. Over the preceding decades the US and to a lesser extent Russia invested widely in advanced conventional platforms. This manifested in 2002 with the Bush administrations' 'new strategic triad' consisting of nuclear *and* conventional offensive capabilities, defensive systems, and secure command, control, and communications.[103] This logic led to the abrogation of the Anti-Ballistic Missile (ABM) Treaty and pushing ahead with various missile defence systems, as well as developing a range of different prompt strike programmes.[104]

Today, the US maintains a significant technological edge when it comes to any aspiration of counterforce targeting of an adversaries' nuclear facilities and systems. In addition to highly accurate *nuclear* weapons, the US also has a very sophisticated conventional precision strike and defence complex. Indeed, Princeton University's Ryan Snyder has argued that the US Tomahawk or joint air-to-surface stand-off missiles (JASSMs) could be used to substitute for nuclear attacks on missile silos—calculating that they have both the accuracy and the blast power.[105] The nascent layered integrated air and missile defence system may well be able to defeat limited missile or air attacks (the plot from the recent Netflix film *House of Dynamite* notwithstanding ...), and the US navy can probably track and sink most adversary nuclear-armed submarines. Finding and destroying mobile missiles may be more challenging—though not impossible.[106] If President Trump's 'Golden Dome' vision were to become a reality (albeit, from the current vantage point, unlikely) this would provide a considerable challenge to *any* would-be attacker against the US.

But the best contemporary example of the non-nuclear pathway to nuclear security is South Korea, and the development of the so-called 'kill chain' and 'three axis' strategy. This involves developing the ability to pre-emptively carry out conventional precision strikes against North Korean nuclear systems and command nodes, and 'decapitating' the North Korean leadership, deploying a multilayered missile defence system, and

[103] See Donald H. Rumsfeld, 'Foreword', Nuclear Posture Review, Secretary of Defense, Project on Govenrment Secrecy, Federation of American Scientists, 2002, https://sgp.fas.org/news/2002/01/npr-foreword.pdf.

[104] James Acton, *Silver bullet: Asking the right questions about conventional prompt global strike* (Washington, DC: Carnegie Endowment for International Peace, 2013).

[105] Ryan Snyder, 'Assessing the lethality of conventional weapons against strategic missile silos in the United States, Russia, and China', *Science & Global Security* 32:1–3 (2024), pp. 105–73.

[106] Thomas MacDonald, 'Tracking mobile missiles', *Journal of Strategic Studies* 48:2 (2025), pp. 297–333.

threatening large-scale conventional punishment in response to a nuclear attack. To achieve this, the South Korean military deploys a *smorgasbord* of systems including stealthy F35a aircraft, Hyunmoo-2 ballistic, and Hyunmoo-3 cruise missiles, the Thaad, Patriot, L-SAM, M-SAM missile defences systems, as well as early warning satellites and radar. In theory South Korea offers a glimpse into what a non-nuclear deterrence future might look like.

However, South Korea may also highlight the limitations of such an approach to nuclear security. As northeast Asia specialists Ian Bowers and Henrik Stålhane Hiim have argued,

> While the South Korean case illustrates how deeply technological shifts are challenging common assumptions about nuclear strategy and, in particular, the role of conventional weapons, it also demonstrates that dilemmas associated with both counterforce and leadership targeting endure.[107]

Of course, South Korea still also has its alliance with the US and the extended nuclear deterrence guarantee that undergirds its conventional force posture. Moreover, as discussed in the next chapter, the Republic of Korea (ROK) also has certain degree of nuclear latency.

Counterforce targeting is hard, and even harder with conventional weapons, but technological change is creating at least the perception that certain types of weapons could increase the vulnerability of nuclear arsenals. Albeit this does not mean that all nuclear arsenals are becoming vulnerable to conventional attacks, or to the same extent: as Leiber and Press explained in 2020:[108]

> To be clear, nuclear arsenals around the world are not becoming equally vulnerable to attack. Countries that have considerable resources can buck these trends and keep their forces survivable, albeit with considerable cost and effort. Other countries, however-especially those facing wealthy, technologically advanced adversaries-will find it increasingly difficult to secure their arsenals, as guidance systems, sensors, data processing,

[107] Ian Bowers & Henrik Hiim, 'Conventional counterforce dilemmas: South Korea's deterrence strategy and stability on the Korean peninsula', *International Security* 45:3 (Winter 2020/21), p. 9.

[108] Keir Lieber & Daryl Press, 'The new era of counterforce: Technological change and the future of nuclear deterrence', *International Security* 41:4 (Spring 2017), pp. 9–49.

> communications, artificial intelligence, and a host of other products of the computer revolution continue to improve.[109]

A concurrent trend is an expanded role for advanced or strategic non-nuclear weapons in deterrence thinking more broadly. In particular, the Third Nuclear Age is likely to be characterised by the spread of long-range precision missiles (or 'fires') and increasingly effective defences. The mission sets for these technologies are quite diverse, but increasingly conventional weapons are seen as a 'doable' way to destroy high value targets. As Fabian Hoffman at the University of Oslo explains: 'LRS [long-range strike] can create strategic levels effect by fulfilling four distinct strategic functions: (1) counterpopulation (2) strategic interdiction, (3) counterleadership, and (4) counterforce'.[110] It is also possible that highly accurate conventional weapons could be used against TNWs.

There are a number of examples of this dynamic playing out across Europe and Asia. Japan is rapidly expanding its missile defence and conventional strike options, including Aegis and PAC BMD, Tomahawk land-attack and joint-strike missiles, and the hypervelocity gliding projectile (HVGP) and hypersonic cruise missile programmes.[111] Taiwan has developed an extensive arsenal of defensive and offensive missiles, and is pursuing an 'iron dome'-like system in preparation for a large-scale Chinese invasion—what is termed the 'Porcupine Strategy'.[112] China has developed a range of dual capable ballistic, cruise, and hypersonic missiles for sea and land attack, possibly linked to the Beidou satellite system, and is involved in building a limited missile defence system.[113] In India, the Brahmos missile developed with Russia, and the Storm Shadow purchased from the UK-French consortium (which was used as part of Operation Sindoor in 2025) is being augmented with the long-range land attack cruise missile with a range over

[109] Keir Liber & Daryl Press, *The myth of the nuclear revolution: Power politics in the atomic age* (Ithaca, NY: Cornell University Press, 2020), p. 69.

[110] Fabian Hoffman, 'The strategic-level effects of long-range strike weapons: A framework for analysis', *Journal of Strategic Studies* 47:6–7 (2024), pp. 889–990.

[111] E.g. 'Defense of Japan 2022', Digest, Ministry of Defense, Tokyo, 2022, https://www.mod.go.jp/en/publ/w_paper/wp2022/DOJ2022_Digest_EN.pdf.

[112] See, for example, James Timbie & James Ellis, 'A large number of small things: A porcupine strategy for Taiwan', *Texas National Security Review* 5:1 (2021/2), pp. 83–93.

[113] Stephen Chen, 'China fields Golden Dome prototype before the US can come up with a plan', *South China Morning Post*, 30 September 2025, https://www.scmp.com/news/china/science/article/3327224/china-fields-golden-dome-prototype. See also Bruce W. MacDonald & Charles D. Ferguson, 'Chinese strategic missile defense: Will it happen, and what would it mean?', *Arms Control Today* 45:9 (2015), pp. 14–20.

1,000 kilometres. India is also building a layered missile defence,[114] and may be seeking to develop a long-range conventional ICBM capability.[115] While Australia, not least through the 2021 'AUKUS' (Australia-UK-US treaty) agreement with the US and UK, through which it will (in theory) obtain nuclear-powered submarines and work on hypersonic missile technology, is moving further in this direction.[116]

In Europe, Poland is ramping up conventional platforms such as Patriot and the Joint Air-to-Surface Standoff Missile–Extended Range (JASSM-ER).[117] And in 2024, France, Germany, Italy, and Poland (later joined by the UK and Sweden) established the 'European Long-range Strike Approach' (ELSA) to develop cruise missiles with ranges over 1,000 kilometres.[118] The French Arianne Group is purportedly exploring the possibility of developing a conventionally armed medium-range ballistic missile (with a ~1,500 mile range) as well as a missile that can be used at 'very high altitudes'.[119] If these plans all reach fruition Europe will likely be awash with long-range and highly accurate conventional delivery platforms that could destroy high-value targets in western Russia.

What these trends suggest is that in the Third Nuclear Age the question of whether counterforce strikes against nuclear-armed states, and even winning wars against nuclear-armed opponents, as viable strategies is being reopened. The major difference today is that it is increasingly possible to foresee this mission being carried out with *conventional* rather than *nuclear*

[114] According to Prime Minister Modi, 'all important places in India, including strategic areas as well as civilian ones like hospitals, railways and centres of faith, will be given 'complete security cover' under it by 203.' See Rajat Pandit, 'Mission Sudarshan Chakra: Multi-layered air & missile shield integrated with offensive weapons', *Times of India*, 16 August 2025, https://timesofindia.indiatimes.com/india/mission-sudarshan-chakra-multi-layered-air-missile-shield-integrated-with-offensive-weapons/articleshow/123328590.cms.

[115] Rabia Akhtar, 'The problem with India's new missile gambit', *Asia News*, 21 July 2025, https://asianews.network/the-problem-with-indias-new-missile-gambit/.

[116] Jonathan Caverley, 'AUKUS: When naval procurement sets grand strategy', *International Journal* 78:3 (2023), https://doi.org/10.1177/00207020231195269.

[117] Lukasz Kulesa, 'Operationalizing the "Polish fangs": Poland and long-range precision strike', *The Nonproliferation Review*, 27:1:3 (2020), pp. 49–60. This is part of a special issue on the topic edited by Tom Plant, 'Non-nuclear armed states, precision strike, and nuclear risk', *The Nonproliferation Review*, 27:1–3 (2020), pp. 1–140.

[118] Lotje Boswinkel, 'Europe, deterrence and long-range strike', *War on the Rocks*, 20 March 2025, https://warontherocks.com/2025/03/europe-deterrence-and-long-range-strike/. As the author rather memorably put it: 'It turns out Europe's favorite American import isn't fast food or Netflix. It's missiles, and Europe needs to start making its own.

[119] See Zuzanna Gwadera, 'A French medium-range ballistic missile and "very high altitude" ambitions', *Missile Dialogue Initiative*, 21 July 2025, https://www.iiss.org/online-analysis/missile-dialogue-initiative/2025/07/a-french-medium-range-ballistic-missile-and-very-high-altitude-ambitions/.

weapons. As discussed in the Conclusion this has several implications for the wider stability of the nuclear order, it also gives a glimpse into a possible non-nuclear deterrence future.[120] In sum, the Third Nuclear Age seems likely to be characterised by far greater use of advanced conventional platforms for compellance, escalation management, and deterrence.

Conclusion: Responsibility, entanglement, and stability

In 1980, the doyen of English School International Relations Theory, Hedley Bull wrote that the US and the Soviet Union had forfeited their claims to be seen as responsible managers of the global nuclear order.[121] For Bull, great power—especially that which came with nuclear armament—created special responsibilities in global politics. Fast forward 45 years and the idea of behaving as a 'responsible nuclear power' is back. Leaders in France, the UK, and India in particular have been at pains to present themselves as 'responsible custodians' of nuclear weapons through adherence to global norms and agreements, emphasising the political not military role of nuclear weapons, preserving proper civilian control over nuclear forces, and maintaining a strictly minimum nuclear stockpile.[122] But at the same time, there has been a return to what Bull might term *irresponsible* nuclear behaviour: expanding nuclear stockpiles, the return of battlefield nuclear weapons and nuclear

[120] Futter & Zala, 'Strategic non-nuclear weapons'.

[121] Hedley Bull, 'The great irresponsibles? The United States, the Soviet Union, and world order', *International Journal* 35:3 (1980), pp. 437–47.

Sebastian Brixey-Williams and Nicholas Wheeler have even made the case for a 'responsibility turn' in how we think about nuclear weapons. See 'Nuclear responsibilities: A new approach for thinking and talking about nuclear weapons', *The British American Security Information Council and the Institute for Conflict Cooperation and Security*, November 2020, https://basicint.org/wp-content/uploads/2020/11/Brixey-Williams-and-Wheeler-Nuclear-Responsibilities-A-New-Approach-for-Thinking-and-Talking-about-Nuclear-Weapons-2020.pdf

[122] E.g. 'UK statement on Pillar 1', Second Preparatory Committee for the 2026 Review Conference to the Treaty on the Non-Proliferation of Nuclear Weapons, Foreign, Commonwealth and Development Office, UK Government, 22 August 2024, https://www.gov.uk/government/speeches/second-preparatory-committee-for-the-2026-review-conference-to-the-treaty-on-the-non-proliferation-of-nuclear-weapons-uk-statement-on-pillar-1.

'As a nuclear weapon state, the UK takes its responsibilities seriously. We remain committed to working with others—including those who may threaten our interests and security—to build stability, transparency, and better mutual understanding, and to identify and implement effective measures which will ultimately shape the conditions for achieving a world free of nuclear weapons'.

At the 2025 Non-Proliferation Treaty (NPT) prep com, France self-identified as a 'responsible nuclear weapon state' in its national report.

use scenarios, and dangerous and incendiary nuclear rhetoric. Moreover, technological change has increasingly blurred the distinction between nuclear and non-nuclear options, which in turn may be reopening the scope for wars between or involving nuclear-armed states and their allies. The result is that we seem a long way from the 1985 'Reagan-Gorbachev' mantra that 'A nuclear war cannot be won and must never be fought'.[123]

One of the major concerns about this unfolding deterrence context where strategies become 'more integrated' and perhaps, as the Carnegie Endowments' James Acton has explained in great depth, 'entangled',[124] secure second-strike forces become more vulnerable and the use of force returns to international statecraft, the delicate balance that underpins nuclear stability may not hold. At least three different visions of the deterrence future are in play: one where all actors continue to adhere to the theory of the nuclear revolution—and nuclear weapons are broadly seen as stabilising nuclear relationships; one where certain actors view limited nuclear use or threats as having utility; and one where elites see the possibility of technological transformation as a way to deter, compel, and coerce through non-nuclear means. The result is that both deterrence and strategic stability in the Third Nuclear Age are likely to become more complex and contested. Indeed, in the world of 2025, none of the traditional metrics of strategic stability appear to be in good order in any of the key nuclear relationships: the absence of incentives to use nuclear weapons or engage in an arms race; the absence of armed conflict; or a peaceful security environment.[125]

[123] The full paragraph reads: 'The sides, having discussed key security issues, and conscious of the special responsibility of the USSR and the U.S. for maintaining peace, have agreed that a nuclear war cannot be won and must never be fought. Recognizing that any conflict between the USSR and the U.S. could have catastrophic consequences, they emphasized the importance of preventing any war between them, whether nuclear or conventional. They will not seek to achieve military superiority'. 'Joint Soviet-United States statement', Summit meeting in Geneva, 21 November 1985, https://www.reaganlibrary.gov/archives/speech/joint-soviet-united-states-statement-summit-meeting-geneva.

[124] James Acton, 'Escalation through entanglement: How the vulnerability of command-and-control systems raises the risk of inadvertent nuclear war', *International Security* 43:1 (2018), pp. 56–99. Also, his forthcoming book. See also Nathan McQuarrie, 'Why China has not acted on Western warnings to "disentangle" conventional and nuclear missile capabilities', *Bulletin of the Atomic Scientists*, 22 August 2025, https://thebulletin.org/2025/08/why-china-has-not-acted-on-western-warnings-to-disentangle-conventional-and-nuclear-missile-capabilities/?utm_source=ActiveCampaign&utm_medium=email&utm_content=CDC%20in%20turmoil&utm_campaign=20250828%20Thursday%20Newsletter%20%28Copy%29&vgo_ee=MuXNQnHCmXBKpNt4fDq8PWIt5Osnqhj4gKGDWngjDzEoDWPCLpf6wrE%3D%3AmRLOEl2kT0BFG%2FgbnJ1UbAhxerVQso4W.

[125] James M. Acton, 'Reclaiming strategic stability', in Elbridge A. Colby & Michael S. Gerson (eds.), *Strategic stability: Contending interpretations* (Washington, DC: Strategic Studies Institute and US Army War College Press: February 2013), pp. 117–18.

In a 2025 *Foreign Affairs* essay, Carter Malkasian and Zachary Constantino warned that a dangerous trend had re-emerged in recent years whereby conventional attacks between and against nuclear-armed states was becoming more prominent.[126] The most notable examples are Ukrainian drone attacks against a Russian early warning radar in 2024[127] and against an aircraft base housing strategic bombers a year later;[128] several waves of Iranian missile attacks against Israel, and the short but bloody war that erupted between India and Pakistan in 2025. Each may provide a glimpse into what may be an increased likelihood of conventional conflict under the nuclear shadow in the Third Nuclear Age. This in turn may test Glen Snyder's theory of the stability-instability paradox:[129] a perception of stability at the strategic nuclear level that leads to more instability, adventurism, and risk taking.

[126] Carter Malkasian & Zachary Constantino, 'Nuclear powers, conventional wars: The dangerous erosion of deterrence', *Foreign Affairs*, 17 July 2025, https://www.foreignaffairs.com/united-states/nuclear-powers-conventional-wars?utm_medium=newsletters&utm_source=twofa&utm_campaign=Is%20America%20Breaking%20the%20Global%20Economy?&utm_content=20250718&utm_term=EWZZZ005ZX.

[127] Xiaodon Liang, 'Ukraine strikes Russia early-warning radars', *Arms Control Today*, July/August 2024, https://www.armscontrol.org/act/2024-07/news/ukraine-strikes-russian-early-warning-radars.

[128] Jake Epstein, 'Ukraine struck Russia's warplanes at a base its fighter-bombers flee to when other airfields are in danger, intel says', *Business Insider*, 1 July 2025, https://www.businessinsider.com/ukraine-attacked-russian-jets-at-base-fighter-bombers-disperse-to-2025-7.

[129] Glen Snyder, 'The balance of power and the balance of terror', in Paul Seabury (ed.), *The Balance of Power* (San Francisco, CA: Chandler, 1965).

4
To great power competition and beyond ...

Introduction: Polarity, stability, and nuclear politics

Nuclear politics and nuclear order since 1945 have been fundamentally shaped by polarity and by the distribution of material and ideational power within the international system. And, to put it bluntly, the nuclear era has therefore been dominated by the United States' (US) emergence as the most powerful actor in global politics after the Second World War. While the configuration of power relations has shifted during this period, the nuclear age has, so far, fundamentally been the *American* nuclear age. This has unfolded in two parts: the structural *bipolarity* of the First Nuclear Age, where concurrently a balance of terror between the superpowers created a certain stability at the nuclear level, and Washington moved ahead with the construction of the institutions of global order. This was followed by the *unipolar* Second Nuclear Age, which was characterised by what former French Foreign Minister Hubert Védrine described as US 'hyperpuissance' or 'hyperpower'.[1] This was a period where US dominance entrenched and expanded the norms of global nuclear politics and the nature of international order. With hindsight, both structures were believed to have facilitated a certain 'stability' in an otherwise anarchic nuclear system.

The argument here is that the Third Nuclear Age will be increasingly *multipolar*—or at least as the Dartmouth College professors, Stephen Brookes and William Wohlforth describe it, reflect 'partial unipolarity'.[2] This will be a world driven by a shift in the global balance of hard and soft influence, shaped by the 'revisionist' agendas of Russia and China, and the rise of India

[1] 'Hubert Védrine, France's voice in the world', *The Economist*, 9 November 2000, https://www.economist.com/europe/2000/11/09/hubert-vedrine-frances-voice-in-the-world.

[2] Stephen G. Brooks & William C. Wohlforth, 'The myth of multipolarity', *Foreign Affairs*, May/June (2023), https://www.foreignaffairs.com/united-states/china-multipolarity-myth. As the authors put it, 'the myth of multipolarity obscures just how much power it [the US] still has.'

The Third Nuclear Age. Andrew Futter, Oxford University Press. © Andrew Futter (2026).
DOI: 10.1093/9780198944126.003.0005

as a major stakeholder.[3] What we are essentially witnessing is a transition from a world of relatively uncontested US unipolarity to one where US influence is increasingly opposed by two and possibly three other nuclear-armed actors seeking to exert power and authority in international politics.[4] Looking ahead it is imaginable that combinations of these actors may join forces at particular moments, at others they may be more antagonistic: this is likely to strain existing geopolitical relationships and regional security complexes. But this shouldn't be misunderstood as a purely systematic and structural change: it has been exacerbated by shifts in US domestic politics and by the influence of a handful of powerful leaders. Taken together these dynamics are shifting the nature of the global *nuclear* order too.

While there is some debate over what constitutes a '*nuclear* great power'—interaction between the US, Russia, China, and India will increasingly shape global politics—albeit likely in different ways. This in turn will impact a dozen or so influential middle powers as they seek to respond to the gradual erosion of relative US primacy, or at least a US political elite that moves beyond a certain post-World War Two consensus on the American role in the world. This doesn't mean that US power and influence over the global *nuclear* order has disappeared, or that we will see a world of three or four evenly matched nuclear-armed great powers, but that it is becoming more bounded and challenged, and, put simply, different to what came before. But this is probably going to have some fairly significant implications for the likelihood of conflict, nuclear proliferation, and the shape of international politics more generally. Indeed, it is conceivable that these geopolitical dynamics may also create new proliferation incentives as the US-led system of extended deterrence, offshore balancing, and non-proliferation perhaps becomes weakened. It may also lessen the ability to combat proliferation as different states challenge established non-proliferation norms, and—notwithstanding the US attacks against Iran in June 2025 and the war of 2026—there is less concerted willingness to enforce them. Ultimately, a drift towards a more contested and multipolar geopolitical order probably means more friction between multiple different visions of the global *nuclear* order in the Third Nuclear Age. It is also likely that nuclear weapons play an ever-greater role in the interactions between the great powers.

[3] For a discussion of this, see Benjamin Zala (ed.), *National perspectives on a multipolar order: Interrogating the global power transition* (Manchester: Manchester University Press, 2021).

[4] Campbell Craig, 'American power preponderance and the nuclear revolution', *Review of International Studies* 35:1 (2009), pp. 27–44.

On the one hand, this fluidity may create new opportunities and challenges for stakeholders in the 'post-colonial', non-nuclear-armed world. States from what is often awkwardly termed the 'global south' may have more scope to balance and bandwagon with different major powers, and push their own nuclear interests and air their grievances. On the other hand, the governments of the non-nuclear-armed world may increasingly be forced to choose between different visions of the global political future dominated by the major nuclear-armed players. Overall, current trends probably paint a picture of a weakening of the Western dominance that has characterised much of the nuclear past—and perhaps the slow death of the Western liberal international order built up since 1945, and which came to dominate after 1991. But while the result may be that different visions for the global order are increasingly beginning to clash, there is likely to be greater instability in general, and new nuclear flashpoints may well emerge, this does not necessarily mean that the centrality of nuclear armament to international politics will erode.

While the nuclear world has technically been multipolar in terms of *actors* since 1952 when the United Kingdom (UK) tested its first atomic device on the Montebello Islands 30 miles or so off the coast of western Australia, and the concept has been around since the 1960s,[5] the difference today is the *influence* of these actors and their divergent visions for the global order, and perhaps in some cases of the global *nuclear* order. In this way, perhaps bearing more resemblance to the '19th-century concert of Europe', as Austria, France, Prussia, Russia, and Great Britain carved out spheres of influence across the globe, than to the years since. The difference this time is that the major players have sophisticated nuclear capabilities not wooden ships, muskets, and horses.

From the War on Terror to great power competition ...

The view from inside the Washington, DC, beltway has always been central to how global nuclear politics has been conceptualised at any given time since 1945. US thinking and strategy largely defined the shape and scope of the First Nuclear Age and the orthodoxy that arose as a result, and was

[5] It is worth noting that the concept of nuclear multipolarity isn't new: with e.g. Cero Elliot Zoppo, 'Nuclear technology, multipolarity, and international stability', *World Politics* 18:4 (1966), pp. 579–606, and John J. Weltman, 'Managing nuclear multipolarity', *International Security* 6:3 (1981), pp. 182–94, being good examples.

especially dominant in constructing the focus of the Second Nuclear Age on rogue actors and nuclear counterproliferation. This is no different today, as recent thinking in Washington suggests that the new or Third Nuclear Age will be all about 'great power competition'—and the revisionist challenges in Europe and especially Asia. China seems to be the main focus of this forecast, especially given recent concerns about rapid nuclear proliferation to 'peer status' and the possible march towards 'Thucydides trap' as the Harvard historian Graham Allison describes it,[6] and is arguably the only genuine strategic competitor to the US in the medium and long term. But Russia's nuclear arsenal and recent aggression against Ukraine, and growing links with Iran, North Korea, and China, mean that this is increasingly being constructed as a 'two-peer' problem. The role of India as a nuclear-armed great power in waiting is likely to be important to the Third Nuclear Age too, but is so far largely unexplored.[7]

The transition from the Second to the Third Nuclear Age is perhaps best captured by the changing dominant narrative about nuclear threats in the corridors of power in Washington, DC, and to a lesser extent the capital cities of the US' major allies. In particular, the shift in nuclear ages maps very closely with the transition from a focus on rogue proliferation in the 1990s, and the War on Terror that followed the '9/11' terrorist attacks against the US in the 2000s, towards the return of major power competition underpinned by nuclear-backed contestation in the mid-2010s.

As the Cold War ended, US nuclear security discourse became focussed on two 'rogue' nuclear threats: Iraq and North Korea, and then President Bill Clinton and his national security team would spend much of the 1990s pursuing different ways to manage these challenges.[8] The threat from Russia was focussed primarily on loose nuclear material that might find its way into the hands of certain regimes, and about the future job prospects of newly unemployed nuclear scientists and technicians rather than nuclear war.[9] While concern about Chinese nuclear aspirations grew (and particularly

[6] Graham T. Allison, *Destined for war: Can America and China escape Thucydides trap?* (Melbourne: Scribe, 2017). As Allison puts it, 'China military advances are rapidly undercutting America's status as a global hegemon and are forcing US leaders to confront ugly truths about the limits of American power' (p. 20).

[7] One exception being the forthcoming (at the time of writing) book, Happymon Jacob, *India after non-alignment: The quest to become a pole in international politics* (Washington, DC: Georgetown University Press).

[8] Captured by then Secretary of Defense William J. Perry in his memoir, *My journey at the nuclear brink* (Stanford, CA: Stanford University Press, 2015).

[9] E.g. Graham T. Allison, *Avoiding nuclear anarchy: Containing the threat of loose Russian nuclear weapons and fissile material* (Cambridge, MA: MIT Press, 1996).

after the suspected theft of US W88 warhead designs), Beijing was a long way from being seen as a strategic *nuclear* competitor.[10] Even the formal nuclearisation of south Asia in 1998 was seen more as a proliferation challenge than a direct threat to US national security, global dominance, or regional interests.[11]

The dangers posed by rogue regimes and 'undeterrable' actors dominated the US security debate during the presidency of George W. Bush and would be a central component of what would become known as 'the Bush Doctrine' to combat 'the axis of evil'.[12] This would continue throughout the 'noughties' as concerns were compounded by the 2006 North Korean nuclear test, and by the gradual rise of Iran as a concurrent focus of nuclear dangers thereafter.[13] While the belief in the wisdom of pre-emption fell away under President Barack Obama, the risk of nuclear terrorism and proliferation to regimes hostile to the US continued. As President Obama remarked on the release of the 2010 Nuclear Posture Review: '[the report] recognizes that the greatest threat to U.S. and global security is no longer a nuclear exchange between nations, but nuclear terrorism by violent extremists and nuclear proliferation to an increasing number of states'.[14]

Over the next few years, something changed in the Washington bubble (and elsewhere),[15] and focus shifted towards a renewed focus on Russia and China as the major challenges facing the US.[16] This was perhaps captured

[10] This was detailed in the 1999 'Report of the Select Committee on U.S. National Security and Military/Commercial Concerns with the People's Republic of China' known as the 'Cox Report'. See Shirley A. Kan, 'China: Suspected acquisition of US nuclear weapon secrets', Congressional Research Service, updated 1 February 2006, https://sgp.fas.org/crs/nuke/RL30143.pdf#search=%22sandia%20chinese%20computer%20china%22.

[11] See Strobe Talbott, *Engaging India: Diplomacy, democracy and the bomb* (Washington, DC: Brookings Institution, 2004).

[12] E.g. Anthony DiFilippo, 'US policy and the nuclear weapons ambitions of the "Axis of Evil" countries', *New Political* Science 28:1 (2006), pp. 101–12.

[13] Or technically since the secret facilities were uncovered at Arak and Natanz in 2002. See Alireza Jafarzadeh, *The Iran threat: President Ahmadinejad and the coming nuclear crisis* (Basingstoke: Palgrave Macmillan, 2007).

[14] 'Statement by President Barack Obama on the release of the Nuclear Posture Review', The White House Office of the Press Secretary, The Press Office, 6 April 2010, https://obamawhitehouse.archives.gov/the-press-office/statement-president-barack-obama-release-nuclear-posture-review.

[15] Or as Patrick Porter puts it 'the blob'. Patrick Porter, 'Why America's grand strategy has not changed: Power, habit, and the U.S. foreign policy establishment', *International Security* 42:4 (2018) pp. 9–46.

[16] Though some challenges from the War on Terror years remained. For a discussion of this, see Andrew Futter & Benjamin Zala, 'The return of nuclear great power politics (or why we stopped worrying about terrorists and the bomb)', *European Journal of International Security* 10:1 (2025), pp. 133–49.

best by the Nuclear Posture Review released by the Trump administration just eight years later in 2018:

> global threat conditions have worsened markedly since the most recent 2010 NPR, including increasingly explicit threats from potential adversaries. The United States now faces a more diverse and advanced nuclear-threat environment than ever before, with considerable dynamism in potential adversaries' development and deployment programs for nuclear weapons and delivery systems.[17]

This was mirrored in the academy, as influential US scholars and outlets also began writing about 'great power competition'. As two members of the first Trump administration (2016–20), Elbridge Colby and A. Wess Mitchell put it in 2020: 'the United States is gearing up for a new era—one marked not by unchallenged U.S. dominance but by a rising China and a vindictive Russia seeking to undermine U.S. leadership and refashion global politics in their favour'.[18]

It is difficult to pinpoint exactly what led to this change in thinking, not least because a number of different factors appeared to accumulate: the 2014 Russian annexation of Crimea, indirect clashes between US and Russian military forces during the war in Syria, accusations by both US and Russian officials regarding violations of the Intermediate-Range Nuclear Forces (INF) Treaty, the interference by Russia in the 2016 presidential election, the poisoning of a Russian defector in the UK in 2018, and the much-hyped development of 'exotic' Russian nuclear weapons; along with a growing concern about the growth of Chinese conventional military and economic power (especially the Belt and Road Initiative),[19] intellectual property theft, and aggressive expansionist plans especially in Asia, including the 'island building campaign' and the '9 dash line' claims, as well as increased military activity in the Taiwan strait. But the full Russian invasion of Ukraine

[17] 'Nuclear Posture Review', Office of the Secretary of Defense, February 2018, p. 4, https://media.defense.gov/2018/feb/02/2001872877/-1/-1/1/executive-summary.pdf.

See also, Péczeli, A., 'The Trump Administration's Nuclear Posture Review: Back to Great Power Competition', *Journal for Peace and Nuclear Disarmament*, 1:2 (2018), pp. 238–255.

[18] Elbridge A. Colby & A. Wess Mitchell, 'The age of great power competition: How the Trump administration refashioned American strategy', *Foreign Affairs*, January/February 2020, p. 118.

See also, Elbridge Colby, *The strategy of denial: American defense in an age of great power conflict* (Yale University Press, 2021).

[19] See, for example, Fiona Cunningham, *Under the nuclear shadow: China's information age weapons in International Security* (Princeton, NJ: Princeton University Press, 2025).

and the revelation of hundreds of Chinese intercontinental ballistic missile (ICBM) silos and fears about a race to nuclear parity would not emerge until later.

Perhaps a better explanation is that the return to great power nuclear competition had been brewing since the end of the Cold War, and was the result of a political environment fundamentally shaped by the US and the West, and never really accepted in Moscow and Beijing. While the US and Western partners were celebrating an 'end of history' and a 'victory for capitalism and liberal democracy' with the end of the Cold War,[20] this arrangement was not to everyone's liking. As the US political scientist Walter Russell Mead put it in a 2014 essay, 'China, Iran and Russia never bought into the geopolitical settlement that followed the Cold War, and they are making increasingly forceful attempts to overturn it'.[21] The argument being that 'US power is the chief obstacle to achieving their revisionist goals'.[22] While this may not quite be the 'clash of civilizations' that the eminent political scientist Samuel Huntingdon warned of in the early-1990s[23]—at least not yet—by the mid-2010s it was increasingly one defined by 'multipolar struggle', as the University of Birmingham's inimitable Patrick Porter put it,[24] characterised by increased competition for influence and power. But it would be wrong to see this shift as purely structural and systemic, it has also reflected the power and influence of personalities with markedly different agendas: Vladimir Putin returned to the Russian presidency in 2012, Xi Jinping took over the leadership in China in 2013, and Donald Trump won the 2016 (and 2024) US presidential election.

A central component of the shift in thinking about the nature of nuclear threats facing the US and its allies has been the increasing influence of what

[20] Most popularly captured by Francis Fukuyama in *The end of history and the last man* (Washington, DC: Free Press, 1992).

[21] Walter Russell Mead, 'The return of geopolitics: The revenge of the revisionist powers', *Foreign Affairs*, May/June 2014, p. 2. 'The tide of history may be flowing inexorably in the direction of liberal capitalist democracy, and the sun of history may indeed be sinking behind the hills. But even as the shadows lengthen and the first of the stars appears, such figures as Putin still stride the world stage. They will not go gentle into that good night, and they will rage, rage against the dying of the light'.

[22] Russell Mead, 'The return of geopolitics'.

[23] Samuel Huntingdon, 'The clash of civilizations', *Foreign Affairs* 72:3 (1993) pp. 22–49.

[24] Patrick Porter, 'Advice for a dark age: Managing great power competition', *The Washington Quarterly* 42:1 (Spring 2019), p. 7.

As Porter continued, 'A mix of structural pressures and self-fulfilling choices by key players have brought about this dangerous state of affairs ... In the medium term, this shift is probably irreversible'.

has been termed 'the two-peer' or 'three-body' problem. Essentially, the Third Nuclear Age will represent the first time since 1945 that Washington has been confronted by two nuclear-armed strategic competitors in Russia and China with geopolitical agendas different to the US—up until recently, China had not been considered a peer nuclear competitor due to its relatively small nuclear force, and of course for a time it was hoped that Beijing would slowly assimilate into the Western order. As the 2022 Biden administration 'Nuclear Posture Review' explained, this was reflected by the 'ambitious expansion, modernization, and diversification' of the Chinese nuclear capability, at the same time that Russia was 'emphasizing nuclear weapons in its strategy, modernize and expand its nuclear forces, and brandish its nuclear weapons in support of its revisionist security policy'. The net result, the review warned, is that China could field a nuclear force of 'at least 1000 deliverable warheads by the end of the decade'.[25]

The idea that the US is facing two 'revisionist' powers has become a powerful narrative. Though of course, what constitutes 'revisionist' is in the eye of the beholder. Essentially, after several decades where the US could pretty much call the shots in Asia, Europe, and the Middle East (albeit with varying degrees of success), this is increasingly being challenged. The Ukraine war is the most tangible manifestation of this, but the grand narrative is arguably driven more by the fear of Chinese aspirations in Asia, and fears that China will overtake the US as the most powerful state in the international system. For sure, while Russian forces and military might have been depleted in Ukraine, Russia remains a serious geopolitical challenger, not least due to its diverse nuclear force. But China represents a different challenge, both economically and perhaps eventually politically, and in the meantime may well be on the path to undermining US conventional regional hegemony. As former US Secretary of Defense Robert Gates put it,

> Not since the Korean War has the United States had to contend with powerful military rivals in both Europe and Asia. And no one alive can remember a time when an adversary had as much economic, scientific, technological, and military power as China does today.[26]

[25] '2022 Nuclear Posture Review', US National Defense Strategy, MDR, 27 October 2022, p. 4, https://media.defense.gov/2022/Oct/27/2003103845/-1/-1/1/2022-NATIONAL-DEFENSE-STRATEGY-NPR-MDR.pdf.

[26] Robert M. Gates, 'The dysfunctional superpower: Can a divided America deter China and Russia?', *Foreign Affairs* 102 (2023), p. 30.

Thus, it has become a major question of Third Nuclear Age geopolitics whether the US has the 'willingness' or the hardware to deter both, especially if growing links with other 'revisionist' powers are included.

As a 2023 report from a study group convened at the Center for Global Security Research at Lawrence Livermore National Laboratory, put it:

> For the first time in its nuclear history, the United States faces two major power adversaries armed with large and diverse nuclear forces, capable of challenging the United States and its allies in limited regional war fought with conventional forces, and bound together by a hostility to U.S.-led global and regional orders and the resolve to bring about their end.[27]

While officials in the Biden administration and prominent experts made it clear that this didn't automatically require an increase in US deployed *nuclear* forces,[28] pressure has mounted to think how the US can deter two peer competitors—possibly working together. Indeed, in 2023, a bipartisan group of high-level US national security officials was formed to look into the two-peer deterrence problem. As the Executive Summary of the Commission on the Posture Review explained,

> The United States faces a strategic challenge requiring urgent action. Given current threat trajectories, our nation will soon encounter a fundamentally different global setting than it has ever experienced: we will face a world where two nations possess nuclear arsenals on par with our own. In addition, the risk of conflict with these two nuclear peers is increasing. It is an existential challenge for which the United States is ill-prepared, unless its leaders make decisions now to adjust the U.S. strategic posture.[29]

[27] 'China's emergence as a second nuclear peer: Implications for U.S. nuclear deterrence strategy', Center for Global Security Research (CGSR), Lawrence Livermore National Laboratory (LLNL), Spring 2023, https://cgsr.llnl.gov/sites/cgsr/files/2024-08/CGSR_Two_Peer_230314.pdf.

[28] See Charles L. Glaser, James M. Acton, & Steve Fetter, 'The US nuclear arsenal can deter both China and Russia', *Foreign Affairs* 5 October 2023, https://www.foreignaffairs.com/united-states/us-nuclear-arsenal-can-deter-both-china-and-russia.

[29] 'America's strategic posture: The final report of the Congressional Commission on the Strategic Posture of the United States', Executive Summary of the Commission on the Posture Review, p.vii, October 2023, https://www.ida.org/-/media/feature/publications/a/am/americas-strategic-posture/strategic-posture-commission-report.ashx.

The bottom line, according to the commissioners, was that the US needed to explore other *nuclear* options beyond the current modernisation programme.[30]

To put it bluntly, the two-peer nuclear challenge opens up some pretty fundamental questions about US deterrence posture and policy. Specifically, how many and what type of nuclear forces are required to hold what type of target sets vulnerable to deter Russia *and* China and provide credible extended deterrence guarantees. If the US wants to be able to credibly threaten countervalue retaliation (against 'soft' targets such as cities) then current configurations are probably sufficient.[31] If a more ambitious deterrence strategy that involves escalation management options and perhaps even counterforce or damage limitation strikes—i.e. targeting all adversaries' nuclear forces—is required, then additions might be needed. As UK nuclear expert Nicholas Wheeler put it during a workshop in 2025, 'you don't have to be a mathematician to work out that Russian plus Chinese deployed nuclear forces will add up to more than the 1550 that the US is [now was] allowed to deploy under the New START [Strategic Arms Reduction Treaty] agreement.'[32] Compounding the 'two-peer' nuclear challenge is an associated concern that it may not just be the arsenals of Russia and China's 'friendship without limits'[33] that the US needs to deter but also a growing North Korean arsenal and a possible future Iranian capability. The recent Russia-Democratic People's Republic of Korea (DPRK) defence treaty and Iranian support for the Russian war effort in Ukraine have fuelled this concern.

The idea that more US nuclear weapons might be needed seems to have gained some traction under the second Trump administration that took office in early 2025. The approach taken to managing this challenge in turn will feed into the debate about whether continuing to adhere to the limitations previously imposed by the New START treaty continues to make sense, and it will further complicate future nuclear arms control.[34] The fate of the

[30] Discussed in Chapter 3.

[31] This is broadly the point made by Glaser et al. in their article: 'The US nuclear arsenal'.

[32] Albeit this number doesn't include 'tactical' weapons, those in reserve, or the arsenals of the UK and France.

[33] For a discussion of this, see Patricia M. Kim, 'The limits of the no-limits partnership: China and Russia can't be split, but they can be thwarted', *Foreign Affairs* 102 (2023), p. 94.

[34] For example, Eric S. Edelman & Franklin C. Miller, 'No New START: Renewing the US-Russian deal won't solve today's nuclear dilemmas', *Foreign Affairs* 3 June 2025, https://

bilateral arms control project that began in the 1960s is addressed in the next chapter ...

But the Third Nuclear Age will be about more than the clashing geopolitical visions and deterrence relationships between the US, a nuclear-armed 'wannabe' great power in China, a Russian leadership seeking to revitalise a sense of diminished global status, and what increasingly appears to be at least two quite different visions for the global future. Instead, it will also reflect a broader shift in the distribution of power within the global nuclear system, and moves towards multipolarity more generally,[35] and particularly a new system increasingly structured around *four* rather than three nuclear-armed poles. At the time of writing this a fourth pole is most likely to be India. While perhaps not a *great* nuclear power in the same way as the US, or a great power nuclear challenger like China, decisions taken about global politics by elites in New Delhi will play a key role in how the Third Nuclear Age will unfold.[36]

India is rarely included in the common conceptualisations of great power competition—especially from the Anglo-American framing. But with nearly one-fifth the world's population, a burgeoning economy, an increasingly sophisticated (albeit currently limited) nuclear arsenal, and its transition from post-colonial state to major global actor, the view of the global political landscape and the shape of nuclear order from New Delhi increasingly matters. The Indian position in global nuclear politics has evolved in a way unlike any other state: in the First Nuclear Age Prime Minister Jawaharlal Nehru in particular was a vocal advocate of nuclear disarmament, but Prime Minister Indira Gandhi authorised a 'peaceful nuclear test' known as 'Smiling Budha' in 1974; in the Second Nuclear Age, India became part of the regional proliferation focus after explicitly conducting a nuclear weapon test in 1998, while in the Third Nuclear Age India increasingly seems to be becoming normalised as a 'responsible' and possibly aspiring great 'nuclear' power akin to the five 'recognised' nuclear states under the Non-Proliferation Treaty (NPT).

www.foreignaffairs.com/united-states/no-new-start?s=EDZZZ005ZX&utm_medium=newsletters&utm_source=fatoday&utm_campaign=Israel's%20Dangerous%20Escalation%20in%20Gaza&utm_content=20250603&utm_term=EDZZZ005ZX.

[35] See Futter & Zala, 'The return of nuclear great power politics'.

[36] For a discussion of this, and an argument for why Indian power may be limited, see Ashley J. Tellis, 'India's great-power delusions', *Foreign Affairs*, 17 June 2025, https://www.foreignaffairs.com/india/indias-great-power-delusions.

Officials in both the US and Russia have spent considerable energy trying to court India over the past two decades. The most concrete nuclear manifestation of this was the signing of a civilian cooperation agreement between the US and India in 2006. As the Observer Research Foundation's Harsh Pant noted,

> The Indo-US nuclear pact has virtually rewritten the rules of the global nuclear regime by underlining India's credentials as a responsible nuclear state that should be integrated into the global nuclear order.[37]

This makes India unique in that it can access many nuclear benefits under the NPT without being a party to the treaty (the same is not true for Pakistan). For US strategists, India potentially offers the ideal partner in trying to balance against China. At the same time, the Indian relationship with Russia has continued to mature. The Indian navy leased a Russian nuclear-powered submarine in 2012, and joint work has continued on the Brahmos II hypersonic weapon.[38] That said, policy elites in New Delhi are conscious of what a closer Russia-China relationship could mean given its border issues and adversarial relationship with Beijing. Geopolitically, India therefore sits at the heart of the *four*-party problem, seeking to balance relations with the nuclear great powers, without looking for formal alliances with any of them. India is also central to the other trilateral deterrence relationship between India, China, and Pakistan.[39]

At the time of writing, it seems unlikely that policy elites in New Delhi will seek to rapidly expand the nuclear weapons stockpile to match the other 'great' nuclear-armed powers, even China. India has relatively cordial relations with both the US and Russia, is a member of both the Quadrilateral

[37] Harsh V. Pant, 'The US-India nuclear deal: The beginning of a beautiful relationship?' *Cambridge Review of International Affairs* 20:3 (2007), pp. 455–472.

[38] Daniel Markey & David Brostoff, 'Friends with limits: The future of Russo-Indian defense ties', *War on the Rocks*, 25 April 2025, https://warontherocks.com/2025/04/friends-with-limits-the-future-of-russo-indian-defense-ties/.

[39] E.g. Salman Bashir, 'The China-India-Pakistan nuclear triangle: Consequential choices for Asian security', *Journal for Peace and Nuclear Disarmament* 5:2 (2022), pp. 336–49. See also Ashley J. Tellis, *Striking asymmetries: Nuclear transitions in southern Asia* (Washington, DC: Carnegie Endowment for International Peace, 2022): 'the parallelisms across the two dyads are fascinating: in each case, the weaker states-India vis-a-visa China and Pakistan vis-à-vis India-is far more concerned about the stronger than is true in reverse, yet the stronger entity remains compelled to persistently keep the weaker in its strategic field of view', p. 4.

Security Dialogue (QUAD) and (Brazil, Russia, India, China, and South Africa (BRICS) groupings, and even its rivalry with China is arguably more about status than nuclear competition (notwithstanding recent border clashes ...). Indeed, Modi and Xi held a high-profile meeting on the sidelines of the Shanghai Cooperation Organization in August 2025 suggesting gradually warming relations.[40] India also has strong links to the non-nuclear-armed world through the Non-Aligned Movement (NAM) and has often been seen as a spearhead of protest against established nuclear practices. Although, as Indian scholars Shahid Iqbal and Reshmi Kazi noted in 2025, official Indian support for disarmament and even non-proliferation initiatives has begun to wane.[41]

Since being re-elected in late 2024, Donald Trump has provided further complication to the great power competition question. One the one hand, as Wellesley College's Stacie Goddard has suggested, the great power competition construct may be coming to an end: 'Rather than compete with China and Russia, Trump now wants to work with thcm, seeking deals that, during his first term, would have seemed antithetical to U.S. interests'.[42] This would appear to be borne out by the National Security Strategy released by the Trump administration in November 2025.[43] We could begin to see a possible pathway towards a global system based on 'spheres of influence',[44] itself a manifestation of the changing realities of power. But it might also suggest that the way that Washington seeks to construct the relationship with Russia and China is increasingly not of its own making as different visions clash. Moreover, and while perhaps too early to draw significant conclusions, the diplomatic fallout between the Trump administration and New Delhi over export tariffs and the purchase of Russian oil in 2025 could be a harbinger

[40] See 'Prime minister's bilateral meeting with Chinese President Xi Jinping', Ministry of Foreign Affairs, Government of India, 31 August 2025, https://www.mea.gov.in/press-releases.htm?dtl/40072/Prime_Ministers_bilateral_meeting_with_Chinese_President_Xi_Jinping_August_31_2025.

[41] Shahid Iqbal & Reshmi Kazi, 'The changing contours of India's arms control and disarmament policy', *India Quarterly*, 2025, https://journals.sagepub.com/doi/abs/10.1177/09749284251328231.

[42] Stacie E. Goddard, 'The rise and fall of great-power competition: Trump's new spheres of influence', *Foreign Affairs*, May/June 2025, https://www.foreignaffairs.com/united-states/rise-and-fall-great-power-competition.

[43] 'National Security Strategy of the United States of America', The White House, December 2025, https://www.whitehouse.gov/wp-content/uploads/2025/12/2025-National-Security-Strategy.pdf.

[44] E.g. Peter Rudolf, 'US geopolitics and nuclear deterrence in the era of Great Power competitions', *Political Science Quarterly* 136:1 (2021), pp. 129–53.

of a more complicated relationship between the US and India in years to come.[45]

Nuclear dominos and the death of the West

For the majority of the nuclear age the US has been the most powerful actor in the global system, and the result has been that the US has built, shaped, and regulated global and regional nuclear security architectures over this period. Consequently, any moves away from this dominance, and towards a more multipolar or at least contested global system will have implications for a number of 'middle powers' especially those that have lived in the Western 'liberal orbit' and 'benefitted' from a US extended nuclear deterrence guarantee and security blanket in the past. If we are witnessing, as some have suggested, a 'death of the Western liberal order' led by the US, then it is possible that some of the dynamics that have limited conflict and perhaps nuclear proliferation in the past might shift. In some ways, as the Nuclear Threat Initiative's Eric Brewer noted in 2021, 'the next 30 years may not look like the last 30 ... countries that depend on the United States for security are likely to be more of a proliferation risk than are rogues'.[46]

The idea that the US may be in decline as a global power is something of a well-trodden trope and can be traced back to at least the 1970s.[47] But today there is a feeling that the US may be both increasingly unwilling and perhaps unable to undergird the international nuclear order that has been created since 1945. While some of this might be reflective of the irregular politics of the Trump administration, it is also the product of a US looking increasingly inward rather than outward,[48] and a global system that is becoming more fragmented and contested. These dynamics have led some to suggest that the 'liberal international order' centred around a particular

[45] Kurt M. Campbell & Jake Sullivan, 'The case for a U.S. alliance with India, Washington should draw New Delhi closer, not push it away', *Foreign Affairs*, 4 September 2025, https://www.foreignaffairs.com/united-states/india-alliance-jake-sullivan-kurt-campbell.

[46] Eric Brewer, 'The nuclear proliferation landscape: Is past prologue?' *The Washington Quarterly* 44:2 (2021), pp. 186–7.

As Brewer notes, this may mean that the 2020s and 2030s will more closely resemble the 1970s rather than the 2000s.

[47] E.g. Michael Cox, 'Is the United States in decline—again? An essay', *International Affairs* 83:4 (2007), pp. 643–53.

[48] Condoleezza Rice, 'The perils of isolationism: The world still needs America—and America still needs the world', *Foreign Affairs* 103 (2024), p. 8.

combination of political, social, and economic values and norms, which was built up and presided over by the US across the First and Second Nuclear Ages, is beginning to unravel. As the eminent International Relations scholar John Mearsheimer warned: 'The liberal international order, erected after the Cold War, was crumbling by 2019 ... It was flawed from the start and thus destined to fail'.[49] Six years later, in a February 2025 interview with *Sky News* in the UK, the highly respected British analyst, Michael Clarke warned that: 'The Western world as a construct is dead. It's been sliding fast the last few years, but in recent weeks it's been the final nails in the coffin.'[50]

The 'West' is a phrase usually used to refer to the group of democratic countries consisting of the US and Canada, most nations in Europe, Australia, and New Zealand. Japan and South Korea are sometimes included too.[51] The Western liberal international order refers to the system of beliefs, norms, and institutions built up by the US since 1945, including those designed to manage the risks posed by nuclear weapons that these countries largely adhere to. The concept of a 'Western' 'liberal' international order is of course contested,[52] but it has been central to how global politics has been shaped since the Second World War.

A central plank of the Western liberal project has been managing nuclear proliferation, both through actively working to enforce *non*-proliferation to those actors that might challenge the prevailing order and through a system of extended deterrence guarantees to 'assure' those friends and allies that might feel vulnerable. As the University of Maine's Rebecca Davis Gibbons has chronicled, the history of nuclear non-proliferation has been as much about the US providing incentives not to pursue nuclear weapons as it has been about threatening punishment.[53] All of the countries that fall

[49] John J. Mearsheimer, 'Bound to fail: The rise and fall of the liberal international order', *International Security* 43:4 (2019), pp. 7–50.

[50] 'Ukraine war: Which countries could Putin go for next? Is Europe vulnerable to a nuclear-armed bully? Your questions answered', *Sky News*, 26 February 2025, https://news.sky.com/story/ukraine-war-which-countries-could-putin-go-for-next-is-europe-vulnerable-to-a-nuclear-armed-bully-your-questions-answered-13317497.

[51] For a discussion of this, see William H. McNeill, 'What we mean by the West', *Orbis* 41:4 (1997), pp. 409–34.

[52] John G. Ikenberry, 'The liberal international order and its discontents', *Millennium* 38:3 (2010), pp. 509–21.

As Ikenberry put in the abstract to the article: 'A grand alternative to liberal order does not exist. To put it sharply: the pathway to the future still runs through institutions and relationships created over the last 60 years. American unipolarity will eventually give way to something new. Power and authority will shift in the global system as they have over the centuries. But rival orders will not emerge—even if new leaders will.'

[53] See Rebecca Davis Gibbons, *The hegemon's toolkit: US leadership and the politics of the nuclear nonproliferation regime* (Ithaca, NY: Cornell University Press, 2022).

under the 'Western' framing are covered by the US extended nuclear deterrence guarantees (with the exception of New Zealand[54] and the handful of non-North Atlantic Treaty Organization (NATO) states in Europe[55]). These security agreements have been seen by some as a central mechanism as to why several states that might have 'gone nuclear' have chosen not to: the NATO guarantee in particular was seen as essential to preventing a future west German or Italian bomb in the 1960s,[56] and in the 1970s to bolstering of the security guarantee to South Korea, which thereby played a role in limiting the nuclear aspirations of President Park Chung-hee.[57]

The worry today is that the US may be less committed to undergirding this global nuclear order, bolstering security ties with allies that ameliorate proliferation cravings, and perhaps even enforcing non-proliferation. As the Council on Foreign Relations, Gideon Rose put it:

> Everyone should understand that if the liberal order falls, the nonproliferation regime would fall with it. And the powers rushing for nukes then would be newly orphaned friends of the United States who are no longer convinced they can rely on American security guarantees, and might even have to fear American coercion.[58]

Even if this extreme version does not come to pass, there are fears that the credibility of historical guarantees are weakening, and, as is discussed below, various actors are having to seriously rethink their nuclear security politics for the first time in living memory. In a more competitive world, there are anxieties that the US might need to choose which allies to support, or may be unable—or unwilling—to provide the support previously pledged. While there may be a feeling—or hope—that some of these dynamics are more reflective of President Trump and his desire to reorient US policy

[54] In 1984, the government led by David Lange rejected the extended nuclear deterrence guarantee provided by the US. See 'Taking a nuclear-free policy to the world', New Zealand Foreign Affairs & Trade, https://www.mfat.govt.nz/en/about-us/mfat75/taking-a-nuclear-free-policy-to-the-world.

[55] Austria, Bosnia, Cyprus, Ireland, Malta, Serbia, and Switzerland.

[56] See Beatrice Heuser, *NATO, Britain, France, and the FRG: Nuclear strategies and forces for Europe, 1949–2000* (Basingstoke: Macmillan, 1997); and Leopoldo Nuti, '"Me too, please": Italy and the politics of nuclear weapons, 1945–1975', *Diplomacy & Statecraft* 4:1 (1993), pp. 114–48.

[57] Gabriela Bernal, 'South Korea's first attempt at going nuclear', *The Diplomat*, 15 February 2023, https://thediplomat.com/2023/02/south-koreas-first-attempt-at-going-nuclear/.

[58] Gideon Rose, 'Get ready for the next nuclear age: How Trump might drive proliferation', *Foreign Affairs*, 8 March 2025, https://www.foreignaffairs.com/authors/gideon-rose.

(and get US allies to pay more towards their security[59]), these pressures are arguably as much about longer term structural global geopolitical shifts. A final possible implication of a more multipolar world is that the ability for non-proliferation norms to be enforced may be weakened or at least carried out in very selective ways: ultimately, it raises the difficult question of if not the US, then who leads on combatting proliferation?[60]

As the Italian scholar Ludovica Castelli has shown in great detail,[61] and as was memorably captured in the lyrics of American singer-songwriter Tom Lehrer's 1965 song 'Who's Next',[62] the fears of 'nuclear dominos' and rapid horizontal nuclear proliferation have always been part of a particular imagined nuclear future.[63] But to a large extent, these anxieties did not materialise across the First and Second Nuclear Ages: only 10 states have built the bomb in 80 years, while the vast majority have chosen different pathways to security.[64] But today we are again at an inflection point, and it is entirely possible that a handful more states could seek to join the nuclear club in the Third Nuclear Age as structures and perceived certainties that have governed international order for decades become more frayed.

Geopolitical changes are having a marked effect on alliance politics, the viability of extended nuclear deterrence guarantees, and future calculations about nuclear proliferation. The combination of growing (or at least a perception of growing) Russian and Chinese power (and threats to incumbent regional security orders) as well as a gut feeling that a more capricious political environment may mean the US is less willing than in the past to support its major allies has upset decades-long security thinking in European and northeast Asian capitals. The challenge is playing out differently of course, in part because of the characteristics of distinctive security architectures and

[59] While this isn't a new thing, there has been a notable step up in rhetoric and threats. See, for example, Jen's Stoltenberg, '"I", leaving', Trump said. "There's no reason to be here anymore": Inside the meeting that brought Nato to the brink', *The Guardian*, 4 October 2025, https://www.theguardian.com/world/2025/oct/04/im-leaving-trump-said-theres-no-reason-to-be-here-any-more-inside-the-meeting-that-brought-nato-to-the-brink.

[60] Albeit, as my friend and colleague Benoit Pelopidas put it to me—the US has actually been the source of most of the nuclear proliferation.

[61] See Ludovica Castelli, 'Metaphorising nuclear futures: The case of the nuclear domino metaphor in the US nuclear proliferation discourse', Unpublished PhD thesis, University of Leicester, 2025.

[62] 'Who's next', Tom Lehrer Songs, December 2018, https://tomlehrersongs.com/wp-content/uploads/2018/12/whos-next.pdf.

[63] The classic example being US President John F. Kennedy's warning of 20–25 nuclear-armed states in 1963.

[64] See Benoît Pelopidas, 'The oracles of proliferation: How experts maintain a biased historical reading that limits policy innovation', *Nonproliferation Review* 18:1 (2011), pp. 297–314.

threat environments, but the broader impact could be magnified. At the same time, there is a feeling that the lack of a dominant power to enforce (rogue state) non-proliferation, or possibly temper the counterproliferation vision of the elites in Tel Aviv could impact the nuclear security politics in the Middle East. This has returned us to the idea of 'nuclear dominos', the so-called '*n*-th' country problem,[65] and the world of nuclear proliferation.

The US signed extended nuclear deterrence agreements with South Korea in 1953 as part of the Mutual Defence Treaty, and with Japan through the US-Japan Security Treaty in 1960: declaring that Washington would come to the aid of both countries if they were attacked. These commitments came to include protection under the nuclear umbrella. While the South Korean government flirted with the idea of an independent nuclear capability in the 1970s, and US nuclear weapons were stationed on the peninsular until 1991, this has largely moderated thoughts of proliferation and formed a cornerstone of the regional 'hub and spoke' system across both First and Second Nuclear Ages. In 2023, the US and South Korea signed the Washington Declaration[66] and the US reaffirmed its 'iron-clad' nuclear guarantee to Japan. In the not too distant past there was even talk of trilateral nuclear coordination to meet the challenges in the region.[67]

But serious concerns about the viability of US extended deterrence in Asia have slowly grown, especially as China is seen as becoming more aggressive in the region. At first this was linked to rumours about the Obama administration looking to change US nuclear use doctrine,[68] but was subsequently compounded as a result of remarks by President Trump about allies needing to pay more for their security, and a general weakening of confidence in the US as the security backstop. In a clear example of how words matter—in 2016, in two rather haphazard interviews, Trump alluded to the possibility of Japan and South Korea getting the bomb, which appeared to go against decades of US policy in the region.[69] At the same time, regional nuclear politics are being driven by developments in the DPRK—particularly the

[65] E.g. Fred Iklé, 'Nth countries and disarmament', *Bulletin of the Atomic Scientists* 16:10 (1960), pp. 391–4.

[66] 'Washington Declaration', Republic of Korea Ministry of Foreign Affairs, 28 April 2023, https://www.mofa.go.kr/eng/brd/m_25772/view.do?seq=14&page=1.

[67] E.g. Bee Yun Jo, 'Conventional-nuclear coordination as a way forward for U.S.-ROK-Japan trilateral security cooperation', *Asia Policy* 20:1 (2025), pp. 71–6.

[68] For detailed commentary from the time, see Michael S. Gerson, 'No first use: The next step for U.S. nuclear policy', *International Security* 35:2 (Fall 2010), pp. 7–47.

[69] Ankit Panda, 'Nukes for South Korea and Japan? Donald Trump sees no problem', *The Diplomat*, 27 March 2016, https://thediplomat.com/2016/03/nukes-for-south-korea-and-japan-donald-trump-sees-no-problem/.

fact that since going nuclear in 2006, Pyongyang has expanded its nuclear stockpile. The more recent possibility of 'tactical' nuclear use by the north has raised the stakes in Seoul. At least part of the result of these trajectories is that, today, there is a growing constituency in both countries for an independent nuclear deterrent to combat both North Korea and China.[70]

Both Japan and South Korea are often considered to be 'latent' nuclear-armed states given their advanced nuclear infrastructure. Japan has the ability to enrich and reprocess fissile material, and has sophisticated missile capabilities. While estimates vary, it is possible that Japan could build a nuclear warhead in a few months, and South Korea in a few years (assuming no outside help).[71] This isn't a new phenomenon, but one that has largely been diluted by the sanctity of the US security guarantee in the past two nuclear ages. Taiwan is also considered a latent nuclear state given its nuclear expertise, and it had a clandestine weapons programme in the 1980s. But as Mark Fitzpatrick put it back in 2016: 'Taiwan is the least likely to acquire nuclear weapons today, even though it was once the party that was most intent on this path'.[72] Nevertheless, and while nothing is automatic, this suggests that we are probably nearer today to additional nuclear-armed states in the region than we have been for at least a generation.

The nuclear deterrence and nuclear proliferation dynamic in Europe is slightly different: this is the result of the collective defence component of NATO, whereby the US (as well as the UK) commit their nuclear forces to the defence of the alliance. According to President Emmanuelle Macron, French nuclear weapons have 'a European dimension', but they are nevertheless not formally committed to NATO—or at least France is not part of the NATO Nuclear Planning Group. As part of the NATO agreement, it is widely believed that the US bases 100 or so B61 nuclear gravity bombs in five or six European countries that could be made available to host air forces at a time of war. This is a considerable reduction from the number of US nuclear weapons deployed on the continent during the Cold War (as Chapter 5 explains).

Europeans appear to have been slower to respond to the possibility of a weakening US commitment to their security than Japan and South

[70] E.g. Daniel Sneider, 'Alarmed by Trump, South Korea mulls Japan-style nuclear option', *Asia Times*, 15 March 2025, https://asiatimes.com/2025/03/alarmed-by-trump-south-korea-mulls-japan-style-nuclear-latency/; and Min-hyung Kim, 'Under what conditions would South Korea go nuclear? Seoul's strategic choice on nuclear weapons', *Pacific Focus* 38:3 (2023), pp. 409–31.

[71] For example, Mark Fitzpatrick, *Asia's latent nuclear powers: Japan, South Korea and Taiwan* (London: Routledge for the International Institute for Strategic Studies, 2016).

[72] Fitzpatrick, *Asia's latent nuclear powers*, p. 153.

Korea, but leaked comments about NATO needing to pay for itself in early 2025,[73] concerns that the US might prioritise the Asia-Pacific (read China),[74] and what has been seen as a sometimes capricious response to Russian aggression against Ukraine have fundamentally challenged nuclear security thinking in Europe. This comes at a time when there is genuine fear of a Russian attack against a NATO member state, and hot on the heels of Russian nuclear sabre rattling in Ukraine. As Héloïse Fayet, Ulrich Kühn, and I explained in a 2024 Forum in *Survival*:

> What matters for European security is that, due mainly to the rise of China, US imperial overstretch and rising populism in America, chances are higher today than at most points in recent European history that the continent will see some form of US pullback in the medium term. Depending on the details, Europeans might then have to reassess the role of nuclear weapons in European security.[75]

While there is not the same level of nuclear latency in Europe as there is in northeast Asia, the proliferation challenge is still acute. The main concern is arguably about a reduction of American conventional weapons and support infrastructure, but there are genuine worries about what to do should the US decide to remove its 'tactical' nuclear weapons, or weaken or remove its commitment to Article V of the Atlantic Treaty.[76] In a worst-case scenario, a post-American Europe would be reliant on the ~260 UK warheads committed to NATO and the ~300 French nuclear warheads that aren't. Neither stockpile compares favourably in terms of numbers with the thousands of Russian weapons, and the difference is especially acute in terms of 'tactical' or 'battlefield nuclear weapons'. As the academics Mark Bell and Fabian Hoffman explain:

> With limited arsenals, particularly when it comes to low-yield weapons that can be used to push Russia back rather than escalate, they [the UK and

[73] Including as part of a now infamous leaked Signal chat where Europeans were described as 'freeloaders'. Katya Adler, 'Disdain for Europe in US Signal chat horrifies EU', *BBC*, 25 March 2025, https://www.bbc.co.uk/news/articles/c204vl27n2qo.

[74] Dan Sabbagh, 'US no longer "primarily focused" on Europe's security, says Pete Hegseth', *The Guardian*, 12 February 2025, https://www.theguardian.com/us-news/2025/feb/12/us-no-longer-primarily-focused-on-europes-security-says-pete-hegseth.

[75] Héloïse Fayet, Andrew Futter, & Ulrich Kühn, 'Towards a European nuclear deterrent', *Survival* 66:4 (2024), p. 68.

[76] Albeit Article V is quite vague ... See 'Collective defence and Article 5', NATO, updated 12 November 2025, https://www.nato.int/cps/en/natohq/topics_110496.htm#:~:text=Article%205%20provides%20that%20if,to%20assist%20the%20Ally%20attacked.

> France] also lack the flexibility to counter Russian threats that fall below the level of total annihilation. Both countries have relied on ambiguity in their nuclear strategies, avoiding rigid commitments on the exact conditions of nuclear weapons use to keep adversaries uncertain. This approach may have been effective when their arsenals were solely focused on homeland defense, but it is unlikely to reassure allies that would depend on these nuclear forces for their defense.[77]

Perhaps as a result, in July 2025 the UK and France signed the Northwood Agreement that will 'deepen their nuclear cooperation and coordination'.[78] The same month, the UK signed the Kensington Treaty with Germany, part of which promised close dialogue on nuclear issues.[79] Some other options currently being mooted include an expansion of UK and French nuclear forces, including reconstituting an *independent* British tactical platform;[80] cooperation between major European states on a 'Euro-nuke' project;[81] discussions about how to formalise the French nuclear commitment to Europe, and increased roles for non-nuclear members of NATO. The decision taken in 2025 for the UK to buy US F35a aircraft and join the NATO sharing mission doesn't obviate this problem. Neither does the call by Polish President Andrzej Duda in 2025 for US nuclear weapons to be deployed on its territory.[82] Both continue to involve dependency on the US. It is at least conceivable that Warsaw could consider pursuing an independent nuclear option if the US link were weakened, although it is not clear how this would

[77] Mark S. Bell & Fabian Hoffman, 'Europe's nuclear trilemma', *Foreign Affairs*, 31 March 2025, https://www.foreignaffairs.com/europe/europes-nuclear-trilemma?s=EDZZZ005ZX&utm_medium=newsletters&utm_source=fatoday&utm_campaign=Europe's%20Nuclear%20Trilemma&utm_content=20250331&utm_term=EDZZZ005ZX.

[78] 'UK-France joint nuclear statement', Northwood Declaration, UK Government, 10 July 2025, https://www.gov.uk/government/news/northwood-declaration-10-july-2025-uk-france-joint-nuclear-statement.

[79] 'Treaty between the United Kingdom of Great Britain and Northern Ireland and the Federal Republic of Germany on friendship and bilateral cooperation', UK Government, 17 July 2025, https://www.gov.uk/government/news/treaty-between-the-united-kingdom-of-great-britain-and-northern-ireland-and-the-federal-republic-of-germany-on-friendship-and-bilateral-cooperation.

[80] As Nick Ritchie and John Walker explain, this would be a very difficult undertaking: 'Irreversibility and nuclear disarmament: The case of denuclearising the Royal Air Force', York Irreversible Nuclear Disarmament (IND) Research Report #3, University of York, April 2024, https://www.york.ac.uk/media/politics/documents/yorkindproject/2024%20Report3%20Ritchie%20&%20Walker%20Denuclearising%20the%20RAF.pdf.

[81] Kjolv Egeland & Benoit Pelopidas, 'European nuclear weapons? Zombie debates and nuclear realities', *European Security* 30:2 (2020), pp. 237–58.

[82] Jeremy Bowen, 'Duda: US nuclear weapons in Poland would be a "deterrent" for Russia', *BBC*, 13 March 2025, https://www.bbc.co.uk/news/articles/ce98ym8k89do.

be achieved and how quickly.[83] Germany, Sweden, and Italy all possess technical capabilities associated with nuclear latency (not in the same way as Japan or South Korea), but future proliferation would be on the margins of the debate.[84] Moreover, depending on what happens with the current war in Ukraine, it is possible that leaders in Kiev see nuclear armament as the only way to prevent future Russian aggression.[85]

A final implication of the 'death of the West' narrative, more contested great power geopolitics, and particularly the perception that the US will be less willing to enforce non-proliferation norms will be on the nuclear dynamics in the Middle East. The Middle East has always been at the heart of—largely Western—concerns about proliferation, and proliferation cascades that could occur.[86] There are two parts to this: first is the ongoing challenge presented by Iran, and the fact that as of 2025 Iranian scientists and engineers may have enriched enough uranium to produce several nuclear bombs.[87] The spectre of an 'Iranian bomb' was a centrepiece of the Second Nuclear Age construct, but the full implications of increased Iranian nuclear latency may only really play out in the Third Nuclear Age. Irrespective of whether Iran chooses to overtly weaponise, the possibility is already shaping concerns amongst US regional allies, as are the growing ties between Iran and Russia. But Iran has also become something of a rallying point for other non-Western states keen to defend the right to enrichment and call out what is perceived as US double standards when it comes to nuclear technology diffusion. At the time of writing, it is impossible to say what the medium- or long-term results of Operation Midnight Hammer and the US air force's conventionally armed 'bunker buster' bomb strikes on Fordow, Natanz, and Isfahan will be.[88] Or the of the war that began in February 2026.

[83] Bartosz Glowacki & Aaron Mehta, 'A nuclear Poland? Logistics, political challenges ahead of potential Warsaw deal', *Breaking Defense*, 13 March 2025, https://breakingdefense.com/2025/03/a-nuclear-poland-logistics-political-challenges-ahead-of-potential-warsaw-deal/.

[84] E.g. Alexander K. Bollfrass, 'A new European nuclear deterrent would not be a quick fix', *International Institute for Strategic Studies*, 2 May 2025, https://www.iiss.org/online-analysis/online-analysis/2025/05/a-new-european-nuclear-deterrent-would-not-be-a-quick-fix/.

[85] Of course, John Mearsheimer had argued that Ukraine should be allowed to acquire nuclear weapons in the early 1990s. See 'The case for a Ukrainian nuclear deterrent', *Foreign Affairs* 72:3 (Summer 1993), pp. 50–66.

[86] E.g. Christopher Hobbs & Matthew Moran, *Exploring regional responses to a nuclear Iran: Nuclear dominos?* (Basingstoke: Palgrave, 2014).

[87] Richard Nephew, 'Is a good Iran deal possible? What Washington needs from nuclear negotiations with Tehran', *Foreign Affairs*, 26 May 2025, https://www.foreignaffairs.com/iran/good-iran-deal-possible.

[88] E.g. Joseph Rodger, 'What Operation Midnight Hammer means for the future of Iran's nuclear ambitions', *CSIS*, 23 June 2025, https://www.csis.org/analysis/what-operation-midnight-hammer-means-future-irans-nuclear-ambitions.

One interpretation is that the attacks may not have destroyed all the nuclear material;[89] a second, shared among non-Western states, is that US and Israeli actions were illegal.[90]

The choices made in Tehran will echo across the region, as will the result of the (as I write) ongoing war. The most likely candidate for nuclear proliferation is Saudi Arabia. Crown Prince Mohammad Bin Salman said in an interview in 2018 that: 'Saudi Arabia does not want to acquire any nuclear bomb, but without a doubt if Iran developed a nuclear bomb, we will follow suit as soon as possible'.[91] It is conceivable that Saudi Arabia could move towards a position of nuclear latency—leaders in Riyadh are keen to sign a nuclear deal that will allow control of the nuclear fuel cycle.[92] But to do this, Saudi Arabia will need to sign a different nuclear agreement to the one between the United Arab Emirates and the US as this does not allow for enrichment.[93] Under this agreement (South) Korea Hydro & Nuclear Power has built the nuclear plant, and fuel is purchased from Kazakhstan. The other possible regional proliferation candidate is Turkey. Whilc Turkey is part of nuclear-armed NATO and may still play host to US B61 bombs on its territory, Turkish security politics are complicated. In 2019, President Erdogan made the following remark; 'Some countries have missiles with nuclear warheads, not one or two. But (they tell us) we can't have them. This, I cannot accept'.[94] This naturally led to concerns about proliferation, particularly given the development of nuclear power underway in the country. But the nuclear plants that are being built will be operated by Rosatom, meaning that Ankara has no obvious pathway to an indigenous bomb.[95] But this relationship with Russia feeds into a bigger concern that different states

[89] Mark Fitzpatrick, 'Attacking Iran and tempting fate', *Survival* 67:4 (2025), pp. 7–24.

[90] See Donald Rothwell, 'Why the US strikes on Iran are illegal and can set a troubling precedent', *The Conversation*, 23 June 2025, https://theconversation.com/why-the-us-strikes-on-iran-are-illegal-and-can-set-a-troubling-precedent-259542.

[91] 'Saudi crown prince: If Iran develops nuclear bomb, so will we', *CBS News*, 15 March 2018, https://www.cbsnews.com/news/saudi-crown-prince-mohammed-bin-salman-iran-nuclear-bomb-saudi-arabia/.

[92] One other speculation is that Saudi Arabia might purchase a bomb from Pakistan.

[93] 'The "gold standard" UAE-US partnership for secure nuclear energy', Embassy of the United Arab Emirates, Washington, DC, updated May 2025, https://www.uae-embassy.org/sites/default/files/2025-05/Embassy_One-Pager_Nuclear%20Cooperation_05.12.2025_V3.pdf.

[94] 'Erdogan says its unacceptable that Turkey can't have nuclear weapons', Reuters, 5 September 2019, https://www.reuters.com/article/world/erdogan-says-its-unacceptable-that-turkey-cant-have-nuclear-weapons-idUSKCN1VP2Q8/.

[95] Ali Alkis & Valeriia Gergiieva, 'Why Russia may control Turkey's nuclear energy for the next 80 years', *Bulletin of the Atomic Scientists*, 21 February 2023, https://thebulletin.org/2023/02/why-russia-may-control-turkeys-nuclear-energy-for-the-next-80-years/.

across the region look to Russia and China to provide economic support for civilian nuclear programmes at the expense of Western suppliers, possibly without the attendant proliferation safeguards ...

Two other 'latent' or 'threshold' states are worth mentioning: Canada and Brazil; but while both are theoretically capable (and Canada participated in NATO nuclear sharing until 1983), neither seems likely to pursue the bomb in the near future.[96]

Postcards from the 'global south'

The 'global south' is an awkward label that refers to a hugely diverse group of countries, spread across the globe, with very different nuclear-historical stories, political interests, and security cultures.[97] But what unites this group is that they are not nuclear armed, with a few exceptions, they are all members of nuclear-weapons-free zones, most are formally 'non-aligned' with a great power,[98] and their political and diplomatic elites (periodically) champion nuclear disarmament, are concerned about the breakdown of norms of nuclear restraint, and worry about the challenges of a more competitive nuclear environment. They are also, largely, countries that broke free from European colonialism in the 19th and 20th centuries. Broadly speaking, we can think of the 'global south' as constituting most countries in Africa, South America, and Southeast Asia and members of organisations such as the Association of Southeast Asian Nations (ASEAN), the African Union, and the NAM, with the most prominent nuclear voices coming from the political leaders in Brazil, Egypt, Indonesia, Mexico, Nigeria, and South Africa.

Historically, the 'global south' has played a role in global nuclear politics, but not necessarily one that has had much influence on policy or been able to shape the construction of nuclear order away from the dominant powers. Perhaps the best example is the role of the NAM in what would later become the First Nuclear Age in coordinating member states from across the

[96] Julian Spencer Churchill, 'Bombs away: A Canadian nuclear weapons program to deter Trump is not feasible', Institute for Research on Public Policy, 16 May 2025, https://policyoptions.irpp.org/magazines/may-2025/canadian-nukes/.

[97] Zachariah Mampilly, 'What the "global south" rally means', *Foreign Affairs*, 1 April 2025, https://www.foreignaffairs.com/world/what-global-south-really-means?utm_medium=newsletters&utm_source=fatoday&utm_campaign=The%20Age%20of%20American%20Unilateralism&utm_content=20250416&utm_term=EDZZZ003ZX.

[98] One exception being the Philippines that has a mutual security agreement with the US.

developing world, particularly in pushing for both disarmament and access to nuclear technologies.[99] The influence of what was then often termed 'the third world' on global nuclear politics can be traced back to the founding of the NAM in 1955. As the *communiqué* from the 'Asian-African' or 'Bandung' conference made clear:

> The Conference considered that disarmament and the prohibition of the production, experimentation and use of nuclear and thermo-nuclear weapons of war are imperative to save mankind and civilisation from the fear and prospect of wholesale destruction. It considered that the nations of Asia and Africa assembled here have a duty towards humanity and civilisation to proclaim their support for disarmament and for the prohibition of these weapons and to appeal to nations principally concerned and to world opinion, to bring about such disarmament and prohibition.[100]

Interestingly (at least in hindsight), the conference was sponsored by India and Pakistan, and attended by China—all of whom would later acquire nuclear weapons.[101] At the same time, the *communiqué* also made it clear that the attendees 'emphasized the particular significance of the development of nuclear energy for peaceful purposes'.[102]

Anti-nuclear sentiment would become central to 'third world' or 'developing world' politics, and nuclear disarmament would be increasingly coterminous with a post-colonial desire to assert the rights of the non-nuclear world not to be destroyed. Essentially, one might say, escaping one power over life and death for another. One of the earliest forerunners of this idea was Indian Prime Minister Jawaharlal Nehru. In a 1954 speech he declared, 'We must endeavour with faith and hope to promote all efforts that seek to bring to a halt this drift to what appears to be the menace of total destruction'.[103] Four years later, a letter by President Sukarno of Indonesia (which

[99] For an overview of this, see William Potter & Gaukhar Mukhatzanova, *Nuclear politics and the Non-Aligned Movement* (London: Routledge for the International Institute for Strategic Studies, 2012).

[100] 'Final communiqué of the Asian-African conference of Bandung', 24 April 1955, https://www.cvce.eu/content/publication/1997/10/13/676237bd-72f7-471f-949a-88b6ae513585/publishable_en.pdf.

[101] Japan also attended and would later sign an extended nuclear deterrence agreement with the US.

[102] 'Final communiqué'.

[103] Jawaharlal Nehru, 'Standstill agreement', Statement in the Lok Sabha, 2 April 1954, https://meaindia.nic.in/cdgeneva/?pdf0601?000.

I have been unable to get hold of) to the *New Statesman* was quoted in a speech by British philosopher Bertrand Russell, where the president warned that 'the nations of European descent, including Russia, are claiming the right to exterminate the uncommitted populations of the world as an accidental by-product of their own frenzied lunacy'. According to Russell, he continued:

> We utterly deny you, he says, the right to cause cancer in our children. It is past time for the West, Communist and anti-Communist alike, to draw back from the edge of complete moral bankruptcy. It is explicitly your task to utilize the skill and technique of your science for peaceful purposes.[104]

In a similar vein, Ghanian President Kwame Nkrumah characterised French nuclear testing in Algeria in a 1960 speech as 'nuclear imperialism':

> We in Africa wish to live and develop. We are not freeing ourselves from centuries of imperialism and colonialism only to be maimed and destroyed by nuclear weapons. We do not threaten anyone and we renounce the foul weapons that threaten the very existence of life on this planet. Rather, we put our trust in the awakening conscience of mankind which rejects this primitive barbarism, and believe firmly in positive non-violent action.[105]

Disarmament activism was also central to the discussions at the first Organisation of African Unity conference in 1963. As his Imperial Majesty Haile Selassie, emperor of Ethiopia, put it in his opening remarks: 'We demand an end to nuclear testing and the arms race because these activities, which pose such dreadful threats to man's existence and waste and squander humanity's material heritage are wrong'.[106] Some 35 years later arguably the preeminent leader of 'pan-Africanism', Nelson Mandela warned:

> We must ask the question, which might sound naive to those who have elaborated sophisticated arguments to justify their refusal to eliminate these

[104] Cited in Bertrand Russell, '1958: Only world government can prevent the war nobody can win', *Bulletin of the Atomic Scientists*, 7 December 2020, https://thebulletin.org/premium/2020-12/1958-only-world-government-can-prevent-the-war-nobody-can-win/.

[105] Speech to the April 1960 Conference of African States.

[106] 'Speeches & statements made at the first Organization of African Unity (O.A.U) summit', African Union, May 1963, p. 11, https://au.int/sites/default/files/speeches/38523-sp-oau_summit_may_1963_speeches.pdf.

> terrible and terrifying weapons of mass destruction—why do they need them anyway![107]

In a very different way, Kenyan scholar Ali Masrui was also prominent in academic disarmament discussions. Masrui argued that African states perhaps should acquire nuclear weapons as this was the only way towards nuclear disarmament, what Seifudein Adem describes as a policy of 'proliferate to abolish'.[108] As my colleague Olamide Samuel put it: 'Mazrui wanted to challenge the flawed logic of the continued possession of nuclear weapons limited to an exclusive club of powerful nations'.[109]

Countries in the 'developing' world have also played a key material—and often overlooked—role in nuclear order. A significant proportion of the uranium used in nuclear weapons has come from Africa: in what historian Gabrielle Hecht describes as 'nuclearity'—in fact, it was fissile material from the Shinkolobwe mine in the Congo that fuelled the bomb dropped on Hiroshima in 1945.[110] And today, several developing states (such as Namibia and Niger[111]) supply raw uranium to international fuel markets. Parts of the developing world were also used for nuclear testing, including in Algeria, French Polynesia, the Marshall Islands, and in areas inhabited by indigenous populations in Australia and the United States. Testing was often seen as key component of 'nuclear colonialism'.[112] A desire to challenge the 'nuclear hegemony' of the developed world would become central to many conceptions of post-colonial identity.

It was partly this heritage that drove states from the global south / non-nuclear-armed world first to engage in the Humanitarian Impacts of Nuclear Weapons Initiative in the early 2010s and later to become strong supporters

[107] Address by President Nelson Mandela at 53rd United Nations General Assembly, New York, United States, 21 September 1998, http://www.mandela.gov.za/mandela_speeches/1998/980921_unga.htm.

[108] Seifudein Adem, 'Proliferate to abolish: Mazrui's perspective on nuclear disarmament', *Peace Review* 36:2 (2024), pp. 208–21.

[109] Olamide Samuel, 'How African leadership made the Ban Treaty a reality', *Aljazeera*, 27 October 2020, https://www.aljazeera.com/opinions/2020/10/27/africa-nuclear-weapons-ban-treaty.

[110] Gabrielle Hecht, *Being nuclear: Africans and the global uranium trade* (Cambridge, MA: MIT Press, 2014).

[111] 'World uranium mining production', World Nuclear, https://world-nuclear.org/information-library/nuclear-fuel-cycle/mining-of-uranium/world-uranium-mining-production.aspx.

[112] Exequiel Lacovsky, 'Opposing nuclear weapons testing in the global south: A comparative perspective', *The International Spectator* 58:4 (2023), pp. 73–90.

of the Treaty on the Prohibition of Nuclear Weapons (TPNW) agreed by the United Nations (UN) in 2017. Specifically, it was driven in part by frustrations with the lack of progress on disarmament through the NPT review process, and in part by a desire to amplify claims of *redistributive justice* (in terms of access to nuclear technologies) and *retributive justice* (in terms of compensation for exploitation and harms caused by nuclear testing and resource extraction).[113] Moreover, there was a feeling that the traditional mechanisms to regulate the nuclear order, and in particular to advance the cause of nuclear disarmament—such as the NPT—were suffering a crisis of legitimacy.[114] Tellingly, many states from the 'global south' took a different view on the importance of access to nuclear technologies to many of the states and advocacy groups based in the 'global north'.[115] In many ways this can be traced back to the 'north-south' debates on nuclear energy in the 1970s, and especially to the 'Persepolis Conference' of 1977.[116]

These justice claims feed into broader frustrations across the non-nuclear-armed world about Western / developed world excesses and double standards on the issue of access to nuclear technologies. This is why many global south states have been broadly supportive—or at least not antagonistic—to the Iranian nuclear programme in the recent past, and why Brazil (with Turkey) attempted to broker a 'fuel swap' deal in 2010.[117] More recently, the decision to supply Australia with nuclear-powered submarines under the so-called AUKUS (Australia-UK-US treaty) agreement has also caused problems because Australia will not have to abide by the same restrictions imposed on the majority of other non-nuclear-armed states, and because the agreement will involve the transfer of nuclear material. Many non-aligned states that have signed the Additional Protocol with the International Atomic Energy Agency (IAEA) to allow intrusive inspections of their own nuclear facilities see double standards in the fact that this won't apply to Australia, or indeed to the nuclear weapons facilities operated by

[113] For a discussion of this, see Andrew Futter & Olamide Samuel, 'Accommodating Nutopia: The nuclear ban treaty and the developmental interests of Global South countries', *Review of International Studies* 50:5 (2024), pp. 799–820.

[114] Manuel Herrera, Tanvi Kulkarni, & Vincente Garrido, 'Nuclear non-proliferation and the global south: Understanding divergences and commonalities', *The International Spectator* 58:4 (2023), pp. 1–15.

[115] See Futter & Samuel, 'Accommodating Nutopia.'

[116] See David Holloway & Leopoldo Nuti (eds.), 'The making of the global nuclear order in the 1970s: Issues and controversies' (London: Routledge, 2020).

[117] See Diego Santos Vieira de Jesus, 'Building trust and flexibility: A Brazilian view of the fuel swap with Iran', *The Washington Quarterly* 34:2 (2011), pp. 61–75.

other nuclear-armed states.[118] As has been pointed out to me in numerous off-the-record interviews; there is a perception that non-proliferation has always trumped peaceful uses and disarmament in the Western discourse, with the perceived excesses of the War on Terror being a classic example. Indeed, this might be a reason why Brazil and Argentina agreed their own nuclear monitoring system—the Agencia Brasileño-Argentina de Contabilidad y Control de Materiales Nucleares (ABACC)—in 1991.[119]

But this shouldn't be confused as some unanimous and coordinated view held by all non-nuclear-armed states. For sure, large parts of the non-nuclear-armed world come together at certain points to magnify their impact on nuclear politics: the NAM and the Ban Treaty being the best examples, but influence, goals, and interests vary considerably across the four continents and roughly 150 different governments. Some states focus on challenging a broader Western dominance which they see as being reflected through an unfair nuclear (and global) order, others prioritise nuclear disarmament as a remedy to an unwanted security predicament, some want compensation for previous harms such as testing, while many are probably driven by redistributive economic claims. But a good chunk of global political elites do not see nuclear issues as a priority, or maintain (or are able to maintain) a large diplomatic cadre to engage on these issues.[120]

Brazil is one of the most interesting actors given its current 'Pro-sub' nuclear-propulsion submarine programme, degree of nuclear latency due to the ability to mine and enrich uranium, and decision to regulate its programme outside the IAEA.[121] Other influential voices include: Mexico—whose government played a particularly prominent role in the Humanitarian Impacts Initiative and the TPNW.[122] South Africa, who under

[118] E.g. Felicia Yuwono, 'AUKUS submarine deal a detriment to nuclear commitments', *The Jakarta Post*, 29 September 2021, https://www.thejakartapost.com/academia/2021/09/28/aukus-submarine-deal-a-detriment-to-nuclear-commitments.html.

For a rebuttal of this, see Statement by UK Ambassador Corinne Kitsell on behalf of Australia, the UK and the US to the International Atomic Energy Agency (IAEA) Board of Governors on IAEA safeguards and AUKUS, 11 September 2025, https://www.gov.uk/government/speeches/nuclear-safeguards-aukus-statement-to-the-iaea-board-of-governors-september-2025

[119] A regime described to me by a Brazilian diplomat as, 'friends watching friends'. See https://www.abacc.org.br/wp-content/uploads/2021/08/Paper-30-year-of-ABACC_-Marco_Marzo.pdf.

[120] Or at least as was pointed out to me by a UK Foreign, Commonwealth and Development Office (FCDO) official, many countries do not have the resources to prioritise nuclear issues.

[121] Togzhan Kassenova, *Brazil's nuclear kaleidoscope: An evolving identity* (Washington, DC: Carnegie Endowment for International Peace, 2014); and Matias Spektor, 'The evolution of Brazil's nuclear intentions', *The Nonproliferation Review* 23:5–6 (2016), pp. 635–52.

[122] Cesar Jaramillo, 'Latin America and the quest for nuclear abolition', in Ramesh Thakur (eds.), *The nuclear ban treaty: A transformational reframing of the global nuclear order* (London: Routledge, 2021).

successive African National Congress (ANC) leaders have taken on the role of 'norm entrepreneur' after being the only state to unilaterally disarm in the early 1990s;[123] and Kazakhstan as the world's largest uranium exporter and instigator of a nuclear fuel bank.[124]

The final area where global south nuclear politics will be key is in how different states choose to align with the competing 'poles' of power in the increasingly contested and perhaps multipolar Third Nuclear Age. In the Second Nuclear Age, there was almost an in-built assumption that most would follow or seek to align with the Western, liberal, democratic states led by the US. As president of the International Crisis Group, Comfort Ero puts it:

> after the Cold War ended, most Western policymakers seemed to expect that developing countries would, over time, embrace the Western approach to democracy and globalization. Few Western leaders seemed to worry that non-Western states might bridle at their norms or perceive the international distribution of power as an unjust remnant of the colonial past.[125]

A natural result of this is that the majority of governments and leaders across the world won't necessarily follow Western norms or automatically reach out to Western patrons due to their increased flexibility to chart their own geopolitical path. Thus, while other factors might be driving 'the death of the West', the reduction of Western influence will also be the result of greater 'global south' agency.

In the Third Nuclear Age global south and non-nuclear-armed 'normative' powers will likely have more weight and room to balance the competing poles. In fact, as Vietnamese expert, Huong Le Thu puts it, 'Global south states may even be able to pursue "multi-alignment"'.[126] Or as arguably the leading South American nuclear thinker Matias Spektor explains 'the rising agency of the global South and the expanding geopolitical consciousness among its peoples have fundamentally altered the dynamics of global

[123] See South Africa chapter in Andrew Futter, Ludovica Castelli, Cameron Hunter, Olamide Samuel, & Francesca Silvestri, *The global Third Nuclear Age: Clashing visions for a new era in international politics* (London: Routledge, 2025).

[124] Togzhan Kassenova, 'Kazakhstan and the global nuclear order', *Central Asian Affairs* 1:2 (2014), pp. 273–86.

[125] Comfort Ero, 'The trouble with the "the global south": What the West gets wrong about the rest', *Foreign Affairs*, 1 April 2024, https://www.foreignaffairs.com/world/trouble-global-south.

[126] Huong Le Thu, 'How to survive a great-power competition', *Foreign Affairs* 102:3 (2023), pp. 30–25.

power.'[127] Alternatively, global south states will need to choose sides. The geopolitical pressures of the Third Nuclear Age are particularly acute in regions such as Southeast Asia that sits on a US-China fault line and potential flashpoint, but many states also recognise that the impacts of any nuclear exchange will likely be global.

One major area where this is playing out is in the politics of nuclear energy. As is discussed in more detail in Chapter 6, many governments across the non-Western/ developing world are looking to nuclear energy as part of the transition towards more sustainable ways to power their economies and meet the growing demands of electricity-hungry populations. But unlike in the past, it is Russia and China—rather than Western suppliers—that are dominating the supply of nuclear reactor technology. This is what American scholars Nicholas Miller and Tristan Volpe describe as the 'rise of the autocratic nuclear marketplace.' The growing links between Russia, China and the developing world may also well be a reason why many global south states haven't called Russia out on Ukraine. As Hanna Notte at the Center for Strategic and International Studies explains,

> Facing an intense backlash from Western states over its invasion of Ukraine, Russia has been pursuing a dual strategy: it has undermined non-proliferation efforts and chipped away at trust in legacy institutions, while also leveraging these forums in pursuit of greater alignment with non-Western states. Many states in the Global South have looked for opportunities amid the resulting fragmentation, not necessarily out of support for Russia, but because of a growing frustration over their inability to pursue their own priorities in nuclear forums.[128]

Conclusion: A more contested nuclear order?

In his 2012 book *Perpetual Menace*, the highly respected Scottish scholar William Walker described the global nuclear order as:

> evolving patterns of thought and activity that serve primary goals of world survival, war avoidance and economic development; and the quest for a

[127] Matias Spektor, 'The rise of the non-aligned: Who wins in a multipolar world', *Foreign Affairs* 104:1 (2025), p. 47.
[128] Hanna Notte, 'Russia, the global south and the mechanics of the nuclear order', *Survival* 66:3 (2024), pp. 50–1.

> tolerable accommodation of pronounced differences in the capabilities, practices, rights and obligations of states.[129]

But as we march further into the Third Nuclear Age there is a feeling that this delicate balance of restraint, acceptance, abstinence, and accommodation is under threat. It may even be true that 'Nobody wants the current world order—and all might be considered as revisionists.'[130] As the Oxford University historian Tim Garton Ash put it in May 2025:

> As we pass the 80th anniversary of the end of the second world war in Europe, every day brings further evidence that a remarkably long-lived US-led international order is over. Everyone is now scrambling to work out what might succeed it. A new multi- polar order? Spheres of influence? A worldwide version of the 19th-century Concert of Europe? By far the most plausible answer, however, is a prolonged and dangerous period of global disorder.[131]

Whether retrospectively thinking of nuclear history across the First and Second Nuclear Ages as stable is blinkered, and the result of post hoc rose-tinted spectacles, there is a feeling that a cocktail of geopolitical fluidity is challenging the orthodox view of global nuclear politics that has existed for over three-quarters of a century. At least part of this is a result of changes in the international distribution of power and influence, and especially the decline in influence of 'the West' and perhaps of the US as the leading authority.[132] This comes after a generation where US power seemed omnipotent, and a succession of US leaders that stuck to a bounded grand power narrative. But as Paul Kennedy memorably wrote over three decades ago, imperial overstretch is a key marker of great power decline, and it may well be that the Second Nuclear Age was the zenith of the US and Western idea.[133]

[129] William Walker, *Perpetual menace: Nuclear weapons and international order* (London: Routledge, 2012), p. 12.

[130] Shivshankar Menon, 'Nobody wants the current world order: How all the major powers—even the United States—became revisionists', *Foreign Affairs*, 2 August 2022, https://www.foreignaffairs.com/world/nobody-wants-current-world-order.

[131] Timothy Garton Ash, 'Brace for disorder as the great power sifts begin', *The Financial Times*, 9 May 2025, https://www.ft.com/content/e45091ae-31c7-46b2-95bb-b8197655cd33.

[132] As Hedley Bull suggested in the late 1970s, 'Even in an anarchical society of states, order is possible. But the maintenance of order is more difficult as the number of great powers increases, because there are more interactions to regulate and more potential sources of conflict'. Hedley Bull, *The anarchical society: A study of order in world politics* (Basingstoke: Macmillan, 1977), p. 66.

[133] Paul M. Kennedy, *The rise and fall of the great powers: Economic change and military conflict from 1500 to 2000* (New York, Vintage Books, 1987).

Sadly, this will probably matter a lot for nuclear politics. First, it is likely to mean greater friction between the nuclear-armed states, especially the US, China, Russia, and maybe India. Second, it is likely that non-proliferation norms will become less enforceable, or enforced in an increasingly ad hoc, extra-legal manner. And third, we may see a new era of nuclear proliferation by US allies. As an all-star cast of American experts wrote in 2025: 'Allies with nuclear weapons will complicate America's ability to exercise its power. They're more likely to chart independent, possibly oppositional political and economic policies. And perhaps counterintuitively, they might make it more likely that the United States gets dragged into a nuclear crisis or war.'[134]

As is discussed in the next chapter, what is increasingly apparent is that two very different visions of the nuclear future are taking root: one where nuclear weapons remain central in an unstable and contested world, and one where nuclear disarmament has become even more necessary. In the Third Nuclear Age these visions have become increasingly antagonistic.

[134] Ankit Panda, Vipin Narang, & Pranay Vaddi, 'Nuclear proliferation will haunt "America first"', *War on the Rocks*, 10 March 2025, https://warontherocks.com/2025/03/nuclear-proliferation-will-haunt-america-first/.

5

Taming Einstein's monsters

Introduction: The breakdown of consensus

In the mid-1990s, British Telecom ran a number of TV adverts featuring the actor Bob Hoskins (of *Who Framed Roger Rabbit* fame), in which he would finish a scene with the epithet 'it's good to talk'.[1,2] As a young teenager in a house with one 14-inch TV with 'just' four channels, for some reason this advert has always stuck in my mind. While clearly a marketing ploy aiming to boost revenue and the company's reputation, the same soundbite nevertheless taps into a pretty fundamental part of our nuclear world. It's far better to talk, even with those you disagree with or dislike, rather than run the risk of nuclear war.[3]

Dialogue, a rough consensus, and even in some cases interpersonal trust[4] have always been central to managing nuclear challenges, avoiding misunderstandings, and finding ways to curb ostensible pressures to build more nuclear weapons, keep them on alert, or to escalate conflicts. In fact, even in the darkest days of the Cold War engagement continued between different stakeholders across United States (US) / Western and Soviet societies—the Apollo-Soyuz joint crewed test space mission in 1975 being just one example.[5] But today, dialogue and interest in cooperation appear to have

[1] The chapter title is borrowed from the title of Martin Amis' book, *Einstein's monsters* (London: Vintage, 1999). As Amis put it, 'We are Einstein's monsters', p. ix.

[2] Robert Bean, '"It's good to talk"—The story behind the campaign', *Campaign*, 16 September 2009, https://www.campaignlive.co.uk/article/its-good-talk-story-behind-campaign/938629.

[3] Or as former United Kingdom (UK) Prime Minister Winston Churchill is supposed to have said during a speech at the White House in June 1954, 'To jaw-jaw is always better than to war-war'. See Susan Ratcliffe (ed.), *Oxford essential quotations* (Oxford: Oxford University Press, 2017).

[4] E.g. Nicholas J. Wheeler, *Trusting enemies: Interpersonal relationships in international conflict* (Oxford: Oxford University Press, 2018).

[5] See Olga Krasnyak, 'The Apollo-Soyuz test project: Construction of an ideal type of science diplomacy', *The Hague Journal of Diplomacy* 13:4 (2018), pp. 410–31. Another good example was the Pugwash conferences on science and world affairs. This became even more pronounced after the Cold War. See, for example, Siegfried Hecker, *Doomed to cooperate: How American and Russian scientists joined forces to avert some of the greatest post-cold war nuclear dangers* (Stanford, CA: Stanford University Press, 2016).

The Third Nuclear Age. Andrew Futter, Oxford University Press. © Andrew Futter (2026).
DOI: 10.1093/9780198944126.003.0006

atrophied—most of the arms control frameworks built up over the past half a century have been abrogated or are under immense strain, and there appears limited appetite amongst the nuclear-armed states for the restraint, arms control, risk reduction, and disarmament that were perceived to characterise the second half of the First Nuclear Age and most of the Second Nuclear Age. At the same time, the vast majority of states that are non-nuclear armed and currently have no desire to change this status are putting ever-more pressure on the nuclear-armed minority. This culminated in the agreement of a Nuclear Ban Treaty at the United Nations (UN) in 2017, which effectively makes nuclear weapons illegal, but has arguably deepened the divide between the nuclear haves and have nots.

What has become increasingly apparent is the (re-)emergence of a major fissure between two very different visions of the nuclear future: one that sees nuclear weapons as a necessary evil in a dangerous world and one that sees the continued existence of nuclear weapons as being a central component *of* this dangerous world. The physical and intellectual space in between these positions seems to be becoming compressed, unwelcoming, and difficult to inhabit. Perhaps what makes this evolving bifurcation so pervasive is that it is a phenomenon manifesting across all levels of society: from the offices of the White House, Kremlin, and Zhongnanhai, via the nuclear diplomatic communities in New York, Vienna, and Geneva, to the intellectual spaces inhabited by academics, professionals, and activists. Indeed, the Third Nuclear Age is likely to be characterised by a growing intellectual divide between a loose alliance of diplomats from non-nuclear-armed states, non-governmental organisations (NGOs), activist groups and disarmament, critical, and peace scholars, and those seeking to manage nuclear and conventional challenges to their security through deterrence, extended deterrence, and enhanced conventional deterrence, where more traditional approaches to addressing nuclear challenges ‘incrementally’ is favoured.

The slow death of nuclear arms control and restraint

A key feature of the emerging nuclear order is a feeling that nuclear arms control is on life support as a concept. Over the past two decades most of the bilateral agreements built up between the US and the Soviet Union / Russia over the past 60 years have unravelled, and there appears limited prospects for nuclear arms control involving the other nuclear-armed states. If anything, as Chapter 3 showed, the tendency is more towards qualitative arms racing, limited nuclear expansion, and a greater acceptance for risk taking. This has led some scholars and commentators to warn of the 'death of arms control': notably, in 2015, Russian expert Alexei Arbatov was already foreseeing 'and end of history for nuclear arms control'.[6] Likewise, in 2019, former US National Security official Linton Brookes wrote about 'the end of arms control'.[7]

The nuclear arms control era probably began with the agreement of the Partial/Limited Test Ban Treaty (PTBT) signed by the US, Soviet Union, and the United Kingdom (UK) in 1963. The treaty outlawed nuclear testing in the atmosphere, underwater, and in space (but not underground). At the time, the only other state to have tested a nuclear device was France in 1960—China would test its first device in 1964. The treaty brought to an end nearly two decades of increasingly larger nuclear explosions, culminating in the massive Soviet 'Tsar Bomba' test in 1961, with an estimated 50 metric ton yield, and the US 'Starfish Prime' high-altitude test in 1962, which disabled some of the first satellites in orbit.[8] By this time, as is discussed in more depth in Chapter 6, the magnitude and regularity of testing (approximately 500 explosive tests in 1963 alone) had raised global consciousness about the effects of testing on the atmosphere, humans, and the planet.

The signing of the PTBT began a slow process of mostly bilateral nuclear arms control between the US and the Soviet Union that would take off with the Strategic Arms Limitation Treaty (SALT) I and the Anti-Ballistic Missile (ABM) Treaty in 1972 (freezing nuclear arsenals and prohibiting nationwide missile defences).[9] These agreements marked the start of the process to *limit* the nuclear arms race. A decade later the first steps were made to *reduce* these nuclear arsenals with the 1987 Intermediate-Range Nuclear Forces (INF) Treaty (removing entire classes of nuclear and non-nuclear missiles) and the 1991 Strategic Arms Reduction Treaty (START) I and 1993 START II, which made drastic cuts to deployed weapons numbers and included comprehensive verification measures.[10]

[6] Alexei Arbatov, 'An unnoticed crisis: The end of history for nuclear arms control', Carnegie Moscow Centre, June 2015, https://carnegie-production-assets.s3.amazonaws.com/static/files/CP_Arbatov2015_n_web_Eng.pdf.

[7] Linton F. Brooks, 'The end of arms control?' *Daedalus* 149:2 (2020), pp. 84–100.

[8] Gilbert King, 'Going nuclear over the Pacific', *Smithsonian Magazine*, 15 August 2012, https://www.smithsonianmag.com/history/going-nuclear-over-the-pacific-24428997/.

[9] E.g. James Cameron, *The double game: The demise of America's first missile defense system and the rise of strategic arms limitation* (Oxford: Oxford University Press, 2018).

[10] START I limited each party to 6,000 warheads and 1,600 strategic launchers; START II (which never actually came into force) would have reduced warhead numbers to 3,000–3,500.

As the world moved towards a Second Nuclear Age, the 1991 Presidential Nuclear Initiatives announced by George H.W. Bush withdrew all US ground-launched short-range weapons deployed overseas—including those based on the territory of North Atlantic Treaty Organization (NATO) allies in Europe and on the Korean peninsula, and removed tactical nuclear weapons from surface ships, attack submarines, and land-based naval aircraft.[11] Then Soviet President Mikhail Gorbachev responded by eliminating all nuclear artillery munitions, removing nuclear warheads for short-range missiles, digging up nuclear land mines, and withdrawing all tactical nuclear weapons from surface ships and multipurpose submarines.[12] While not part of an arms control agreement as such, the UK retired free-fall nuclear bombs and France retired nuclear land-based missiles in the 1990s.

However, it is important to note that nuclear arms control throughout this period remained overwhelmingly bilateral and focussed on the two main First Nuclear Age protagonists. This was partly because the US and the Soviet Union / Russia accounted for the vast number of nuclear warheads in existence (on average well over 90%), but also because relatively little progress was being made between other nuclear-armed actors. The UK, China, and France have never signed a nuclear arms control treaty outside of the PTBT (just the UK) and the Comprehensive Test Ban Treaty (which China hasn't ratified). While, since the Middle East effectively went nuclear in the 1970s, south Asia in the 1980s, and northeast Asia in the 2000s, neither have Israel, India, Pakistan, or North Korea signed a binding nuclear arms control agreement.

So, one of the important things to remember about the Third Nuclear Age nuclear 'arms control crisis' is that this is actually a very limited—albeit very important—phenomenon because there wasn't that much to start with. Nuclear arms control may be on life support in the US-Russia context, but in other nuclear relationships it was never born.

The erosion of the suite of US-Russian bilateral nuclear arms control frameworks can probably be traced back to the middle of the Second Nuclear Age, and particularly to the post-9/11 era as the US sought to take

[11] President George H.W. Bush, 'Address to the nation on reducing United States and Soviet nuclear weapons', The American Presidency Project, 27 September 1991, https://www.presidency.ucsb.edu/documents/address-the-nation-reducing-united-states-and-soviet-nuclear-weapons.

[12] Fred Hiatt, 'Gorbachev pledges wide-ranging nuclear cuts', *The Washington Post*, 6 October 1991, https://www.washingtonpost.com/archive/politics/1991/10/06/gorbachev-pledges-wide-ranging-nuclear-cuts/cb8fcb72-0487-4612-9238-e9a6cba71316/.

a different approach towards managing nuclear threats. The most obvious manifestation was the decision taken by the George W. Bush administration to abrogate the Anti-Ballistic Missile (ABM) Treaty in 2002.[13] The ABM Treaty, which was signed as part of the SALT agreements in 1972, had significantly curtailed the development of anti-missile systems, which in turn many believed helped lay the foundations for managing the nuclear arms race. As a result, for some, the ABM Treaty was seen as 'the cornerstone of strategic stability'.[14] However, thirty years later this treaty was seen as getting in the way of protecting against the major Second Nuclear Age challenges facing the US: namely rogue states and terrorists. As President Bush said at the time:

> I have concluded that the ABM Treaty hinders our government's ability to develop ways to protect our people from future terrorist or rogue state missile attacks ... protecting the American people is my highest priority as Commander in Chief, and I cannot allow the US to remain part of a Treaty that prevents us from developing effective defenses.[15]

A key component of this conceptualisation was that Russia was no longer considered a major nuclear threat, as Bush explained in the same speech: 'this is not a day for looking back. This is a day for looking forward with hope, and anticipation of greater prosperity and peace for Russians, for Americans and for the entire world'.[16] While the US and Russia agreed the Strategic Offensive Arms Reduction Treaty (SORT) in 2002,[17] reducing the number of deployed strategic nuclear warheads to between 1,700 and 2,200, since this time, the decision to abrogate the ABM Treaty has regularly been cited by Russia (and China) as a reason driving forward their own nuclear programmes.

[13] For a detailed overview of this decision, see Andrew Futter, *Ballistic missile defence and US national security policy: Normalisation and acceptance after the Cold War* (London: Routledge, 2012)

[14] As described by Russian Foreign Minister Igor Ivanov in 'The missile-defense mistake: Undermining strategic stability and the ABM Treaty', *Foreign Affairs* 79:5 (2000), pp. 15–20. The view was also widely held among Democrats in the US, see Futter, *Ballistic missile defense.*

[15] George W. Bush, 'President discusses missile defense', White House Office of the Press Secretary, 13 December 2001, https://georgewbush-whitehouse.archives.gov/news/releases/2001/12/20011213-4.html.

[16] Bush, 'President discusses missile defense'.

[17] Also known as 'Moscow Treaty', https://media.nti.org/documents/sort_moscow_treaty.pdf. A popular criticism at the time was that this was a 'sort of a treaty' because it didn't contain any verification mechanisms.

Eight years later, the New START treaty of 2010 appeared to represent a significant moment for US-Russian arms control, and even 'recognised the link with missile defences'.[18] The treaty limited both parties to 1,550 deployed strategic nuclear warheads (although importantly it didn't address tactical weapons or weapons held in reserve). A sceptic might argue that the New START treaty effectively codified the status quo and what both the US and Russia could and wanted to deploy at the time. Nevertheless, it allowed for the continuation of a monitoring of each signatories' stockpiles and forces in order to verify compliance. While there were proposals to cut further and broader,[19] the New START restrictions probably represented the pinnacle of the process that had begun in 1972, and perhaps the apogee of Second Nuclear Age arms reductions.

The US pursued arms control discussions with North Korea in the 1990s and 2000s and with Iran throughout from the early 2000s onwards. The 'six party talks'[20] which took place with North Korea between 2003 and 2007 ultimately failed, culminating in the nuclear tests of 2006 and 2009.[21] There was more hope for arms control with Iran, the high point being the signing of the Joint Comprehensive Plan of Action (JCPOA, or 'Iran Deal') in 2015, negotiated by the Obama administration and representatives from China, the European Union, France, Germany, Russia, and the UK.[22] Some limited confidence building measures were agreed between India and Pakistan during this time, including a hotline established in 2004, but no formal arms control agreements were made.[23] Moreover, the Kargil War of 1999 and a number of subsequent crises throughout the 2000s helped push

[18] 'New START Treaty', US Department of State, https://www.state.gov/new-start-treaty.

[19] E.g. 'Modernizing U.S. nuclear strategy, force structure and posture', Global Zero, US Nuclear Policy Commission Report, 2012, https://www.globalzero.org/wp-content/uploads/2018/09/gz_us_nuclear_policy_commission_report.pdf. See also Bryan Bender, 'US generals ask cut in nuclear stockpile', *The Boston Globe*, 6 August 2012, https://www.bostonglobe.com/news/nation/2012/08/05/top-air-force-officer-backup-supply-nuclear-warheads-larger-than-needed/mYLOP5NoK47ko7diYXvFdK/story.html.

[20] US, North Korea, China, Russia, South Korea, and Japan.

[21] See, for example, Mike Chinoy, *Meltdown: The inside story of the North Korean nuclear crisis* (New York: St Martin's Press, 2008).

[22] European Union, 'Nuclear agreement—JCPOA: The Joint Comprehensive Plan of Action (JCPOA) and its implementation, nuclear agreement with Iran', External Action, The Diplomatic Service of the European Union, 18 August 2021, https://www.eeas.europa.eu/eeas/nuclear-agreement-—jcpoa_en.

[23] Haleema Saadia, 'Nuclear risk reduction between India and Pakistan', *Stimson Centre*, 22 May 2024, https://www.stimson.org/2024/nuclear-risk-reduction-between-india-and-pakistan/#:~:text=/00358533.2019.1591767.-,Communication,of%20dialogue%20and%20information%2Dsharing.&text=Mishra%20and%20Ahmed%2C%20"Cooperative%20Measures,Balancing%20and%20Stabilizing%20South%20Asia.

both New Delhi and Islamabad to expand their nuclear capabilities over the proceeding years.

As is discussed in the previous chapter, after the relative high point of the early 2010s, the arms control consensus slowly began to fall apart. In 2017, the Trump administration withdrew from the JCPOA, citing it as 'a poor deal'.[24] In 2019, after long debates about its viability and the extent to which both the US and Russia were adhering to it, the Trump administration withdrew from the 32-year-old INF treaty,[25] in doing so creating the possibility of a return of intermediate-range missiles to Europe (and elsewhere). In 2020, the US withdrew from the Open Skies treaty (which allowed for unarmed overhead surveillance of military activities and deployments)—Russia followed a year later. In 2023, President Vladimir Putin suspended Russian participation in New START and the Russian Duma withdrew ratification for the Comprehensive Test Ban Treaty. And in November 2023 the Russian Duma gave consent to pull out of the Conventional Forces Europe treaty—not a *nuclear* arms control agreement *per se*, but a key component of creating the conditions for nuclear stability. The result was that, the only major bilateral nuclear agreement remaining between the US and Russia was New START, which had been extended in 2021, but which expired in February 2026.

Co-founder of the Stimson Center, Michael Krepon, captured the essence of this story in his magisterial 2021 book *Winning and Losing the Nuclear Peace*:

> The story begins with dashed hopes for abolition, moves on to halting first steps, followed by agreements that broke new ground but could not stop the momentum of the nuclear arms competition. The apogee of success is reached at the end of the Cold War, followed by wilful unravelling of agreements in order to pursue freedom from constraints.[26]

Despite Krepon's calls for a 'revival', the story does not seem to portend well for nuclear arms control in the Third Nuclear Age. It is possible that as of 2026 there will be no formal agreements amongst the nuclear-armed

[24] See, for example, Wendy W. R. Sherman, 'How we got the Iran deal: And why we'll miss it', *Foreign Affairs* 97:5 (2018), pp. 186–98.

[25] Michael R. Pompeo, 'US withdrawal from the INF treaty', Press Statement by Michael R. Pompeo, secretary of state, (2 August 2019), https://2017-2021.state.gov/u-s-withdrawal-from-the-inf-treaty-on-august-2-2019/.

[26] Michael Krepon, *Winning and losing the nuclear peace: The rise, demise and revival of nuclear arms control* (Stanford, CA: Stanford University Press, 2021), p. 1.

states (outside of the commitment towards disarmament for five of these states under the Non-Proliferation Treaty (NPT)—which is itself a problematic issue discussed below). And it is currently hard to see where the next arms control initiatives will come from. While not impossible, it is also difficult to see a return to the formal agreement of the New START treaty between the US and Russia, which means an end to the inspection's regime designed to enhance confidence in compliance (so-called 'trust but verify'). The US appears interested in engaging China in nuclear arms control, but the Chinese politico-military elite seem unwilling until their own nuclear capabilities are closer in number to those of the US and Russia. At the same time, Russian and Chinese officials push back, arguing for the inclusion of the UK and France in any discussions with the US. The likelihood of 'tactical' nuclear weapons being reduced or removed from Europe (a constant theme in arms control discussions and debate in the Second Nuclear Age) has also probably gone for the time being—notwithstanding a unilateral move by the US discussed in the previous chapter). Outside a few confidence building measures, there are no formal bilateral limits on Indian and Pakistani nuclear forces. The same is true of North Korea *vis-à-vis* its immediate neighbours and the US. And there are no nuclear arms control agreements in the Middle East.

So what has caused this breakdown? And who or what is responsible? Part of the reason is arguably technological, particularly the inclusion of new types of non-nuclear weapons and support systems that complicate the bi- or trilateral arms control environment.[27] As discussed in Chapter 2, fears that conventional precision strike weapons, missile defences, and non-kinetic weaponry could be used to undermine nuclear systems is making prospects of nuclear arms control more complicated. Indeed, as discussed in Chapter 3, this is a major reason why the nuclear-armed states are in the process of modernising and in some cases expanding their nuclear forces. But it is also fundamentally political, as different visions for control of the global order have begun to clash. As the respected Russian expert Dmitri Trenin put it:

> the Cold War model of nuclear arms control is inadequate for a world where the relationship between the United States and Russia has undergone fundamental transformation, become hugely asymmetrical, and

[27] For a discussion of this, see Michael Albertson, 'Closing the gap: Aligning arms control concepts with emerging challenges', *Livermore Papers on Global Security*, February 2022, https://cgsr.llnl.gov/sites/cgsr/files/2024-08/CGSR_Livermore_Paper_10_Closing_the_Gap.pdf.

> ceased to be the central axis of global politics. Unsurprisingly, it also shows the immense complexity of organizing an arms control regime with several nuclear powers of different caliber with incompatible strategic cultures and complicated relationships among themselves.[28]

A final knock-on effect from this is that a culture of arms control that became embedded amongst many of the nuclear-armed states over the past generation and the number of people with arms control experience will slowly dissipate. Thus, even if we return to arms control it may be difficult to find the people and organisations to carry it out.

Nevertheless, as pre-eminent American nuclear scholars Morton Halperin and Thomas Schelling explained many decades ago, nuclear arms control is not just about formal, legally binding, verifiable agreements, but about the full gamut of things that might be done to limit nuclear risks:

> Arms control is necessarily thought of as entailing formal agreements, negotiated in detail at diplomatic conferences, embodied in a treaty, and with machinery or institutions for monitoring the agreement. But a more variegated and flexible concept of arms control is necessary—one that recognises that the degree of formality may range from a formal treaty ... through executive agreements, explicitly but informal understandings, tacit understandings, to self-restraint that is consciously contingent on each other's behaviour.[29]

The challenge in the Third Nuclear Age will be finding these areas and building the capacity for when more propitious political circumstances hopefully return: there are ideas of how to address this, but, at the time of writing, little political will.[30] With the following two caveats: (i) arms control doesn't necessarily require cordial relations, but it does require genuine political investment, and this isn't very conspicuous at the moment between the major nuclear powers; (ii) thinking clearly about the question of arms control *of what* and *for what purpose*? This doesn't automatically mean reductions in nuclear capabilities, it could of course mean building up—or at least threatening to—it is important to note that arms control shouldn't

[28] Dmitri Trenin, 'Stability amid strategic deregulation: Managing the end of nuclear arms control', *The Washington Quarterly*, 43:3 (2020), p. 167.

[29] Thomas Schelling & Morton Halperin, *Strategy and arms control* (Mansfield Center, CT: Martino Publishing, 2014 [1961]), p. 77.

[30] E.g. Ulrich Kuhn & Heather Williams, 'A new approach to arms control: How to safeguard nuclear weapons in an era of Great-Power politics', *Foreign Affairs*, 14 June 2023, https://www.foreignaffairs.com/united-states/new-approach-arms-control.

necessarily be seen as synonymous with moves towards disarmament. Likewise, while 'meaningful' arms control is preferable, even the process and symbolic gestures can be important.

'Orthodoxy' versus the Ban Treaty?

In 1968, then non-aligned Finland became the first state to sign the Treaty on the Non-Proliferation of Nuclear Weapons;[31] an agreement that would more colloquially become known as the 'Non-Proliferation Treaty' or NPT for short. What would become the NPT was the brain-child of Frank Aiken, minister for external affairs for non-nuclear-armed Ireland who had begun the process to broker a wide-ranging multilateral agreement to contain the nuclear genie as early as 1958.[32] While two of the then nuclear-armed states, France and China, and a handful of other states suspected (in some cases correctly!) of pursuing clandestine nuclear programmes chose not to sign right away (Argentina, Brazil, India, Israel, Pakistan, and South Africa), by the mid-1990s the treaty had become almost universal. The NPT effectively froze the global nuclear order in late 1960s aspic—those that were nuclear-armed before 1968 and signed the treaty were legally considered 'nuclear weapons states' (NWS), all those who developed nuclear capabilities after this date could never join the treaty as an NWS. Thus, India and Pakistan have never joined the treaty, and can never be recognised as NWS under its current guise. This state of affairs was described as 'nuclear apartheid' by former advisor to the Indian prime minister, Jaswant Singh in an influential article in *Foreign Affairs* in 1998.[33] North Korea joined the treaty as a non-nuclear weapons state (NNWS) in 1982 and became the only state to leave the treaty in 2003.

While the NPT has ten 'articles', it has often been referred to as having 'three pillars' around which it is based. First, a commitment to the non-proliferation of nuclear weapons to more states beyond the US, Soviet Union / Russia, UK, France, and China; second, a commitment to make

[31] Embassy of Finland, 'Finland an active promoter of the NPT', Embassy of Finland, London, 8 May 2015, https://finlandabroad.fi/web/gbr/current-affairs/-/asset_publisher/4ojVRXvbDRVa/content/finland-an-active-promoter-of-the-npt/384951#:~:text=Finland%20was%20the%20first%20to,State%20party%20to%20the%20NPT.

[32] '60th anniversary of Irish Resolution: A forerunner of the NPT', National Security Archive, https://nsarchive.gwu.edu/briefing-book/nuclear-vault/2018-10-29/60th-anniversary-irish-resolution-forerunner-npt

[33] Jaswant Singh, 'Against nuclear apartheid', *Foreign Affairs* 77:5 (1998), pp. 41–52.

nuclear technologies available to all for *peaceful* purposes; and third, a commitment by the five recognised nuclear powers to work 'in good faith' towards nuclear disarmament.[34] At the original time of signing, at the height of the Cold War, there was general recognition that nuclear weapons would continue to play a role in deterrence, albeit hopefully mitigated by arms control. Thus, a certain 'orthodoxy' of global nuclear order became established, one that consisted of a managed system of deterrence and arms control, restraint, access to peaceful technology, and an open-ended commitment to disarmament.[35] In effect, the challenge of non-proliferation came to dominate, and the status quo became increasingly reified, possibly even as something to be protected. This 'disarmament for non-proliferation and peaceful uses' grand bargain, which would be administered to some extent by the International Atomic Energy Agency,[36] would become central to the construct of a First Nuclear Age.

In 1995, the NPT was extended 'indefinitely',[37] and given the shifting geopolitical context, and the transition from a First to Second Nuclear Age, there was a genuine belief that the disarmament pledge might be realised. This hope was further boosted a year later as the Comprehensive Test Ban Treaty—outlawing all nuclear testing—opened for signature in Geneva,[38] and the International Court of Justice in the Hague ruled that the use of nuclear weapons would 'generally not be in accordance with international humanitarian law'.[39] During this period, global nuclear stockpiles reduced considerably, many nuclear delivery systems were retired, South Africa unilaterally disarmed, the former Soviet republics of Belarus, Kazakhstan, and Ukraine all 'gave up' the nuclear weapons that had been stationed on their territory at the end of the Cold War,[40] and nuclear-weapons-free zones (NWFZs) were established in Mongolia (1992) and Southeast Asia (1997). Much as with bilateral arms control, the first half of the Second Nuclear

[34] See UN Office for Disarmament Affairs Treaty database, https://treaties.unoda.org/t/npt.

[35] See William Walker, *The perpetual menace: Nuclear weapons and international order* (London: Routledge, 2012).

[36] See Elisabeth Roehrlich, *Inspectors for peace: A history of the International Atomic Energy Agency* (Baltimore, MD: Johns Hopkins University Press, 2022).

[37] For a discussion of this, see Tariq Rauf & Rebecca Johnson, 'After the NPT's indefinite extension: The future of the global nonproliferation regime', *The Nonproliferation Review* 3:1 (1995), pp. 28–42.

[38] Treaty booklet, Comprehensive Nuclear-Test-Ban Treaty (CTBT), https://www.ctbto.org/sites/default/files/2023-10/2022_treaty_booklet_E.pdf.

[39] 'Legality of the threat or use of nuclear weapons', Request by the General Assembly, ICJ Reports 1996, https://www.icj-cij.org/case/95.

[40] See Mariana Budjeryn, *Inheriting the bomb: The collapse of the USSR and the nuclear disarmament of Ukraine* (Baltimore, MD: Johns Hopkins University Press, 2022).

Age appeared to represent a golden age of reducing nuclear armaments and nuclear dangers.

But over the next two decades, and despite the NPT becoming almost universal,[41] the rough consensus that had existed up to and including the 1995 decision to extend the treaty indefinitely came under mounting pressure. The main issue revolved around the 'commitment to disarmament'. The five NWS could point to significant cuts in their nuclear stockpiles since their heights during the Cold War and emphasised the disarmament *process*. But many NNWS instead emphasised the lack of genuine willingness towards disarmament as an *end goal*. The NWS spoke of an 'incremental approach to nuclear disarmament',[42] and disarmament when political and security conditions allowed. NNWS increasingly pointed to the fact that the NWS themselves were the problem. The result was what with hindsight came to be seen as the relative nuclear consensus of the First and early Second Nuclear Age slowly began to fall apart.

The first major manifestation of this dissatisfaction was the launch of the New Agenda Coalition in 1998 by representatives from Brazil, Egypt, Ireland, Mexico, New Zealand, and South Africa—calling for 'meaningful progress on disarmament'.[43] But the crunch point was arguably the unsuccessful NPT Review Conference in 2005, after which academics and NGOs in particular began to debate the continued validity of the NPT framework.[44] Partly in response, in 2009 former UK secretary of state for defence, Des Browne, launched 'the P5 process' designed to bring together the nuclear-armed members of the NPT to find ways of lowering nuclear risks (and assuaging NNWS' concerns).[45]

Over subsequent years, and despite the agreement of a 64-point action plan at the 2010 NPT Review Conference—including an unequivocal

[41] Only India, Israel, North Korea, Pakistan, and South Sudan are not parties to the treaty as of 2025. Taiwan signed the treaty in 1968 but was replaced by China in 1971.

[42] See Angela Kane, 'What can the incremental approach to disarmament deliver?', in Bård Steen & Olav Njølstad (eds.), *Nuclear disarmament: A critical assessment* (London: Routledge, 2019).

[43] See Robert Green, 'A fast track to zero nuclear weapons: The middle powers initiative and the new agenda coalition', *Medicine, Conflict and Survival* 16:1 (2000), pp. 24–36.

[44] This was an argument made by Michael Wesley in 2005, albeit his argument was that the main problem was preventing future proliferation rather than the lack of progress on disarmament. Michael Wesley, 'It's time to scrap the NPT', *Australian Journal of International Affairs* 59:3 (2005), pp. 283–99. For the opposing viewpoint, see Marianne Hanson, 'The future of the NPT', *Australian Journal of International Affairs* 59.3 (2005), pp. 301–16.

[45] Maximilian Hoell, 'The P5 process: Ten years on', *European Leadership Network*, September 2019, http://www.jstor.org/stable/resrep22113.

commitment to nuclear abolition[46]—belief in the NWS to fulfil their disarmament pledge increasingly eroded. The result, according to Austrian diplomat and instrumental figure in the process to agree the Treaty on the Prohibition of Nuclear Weapons (TPNW) Alexander Kmentt, was that:

> a small group of leading disarmament diplomats, academics and NGO experts began to devise a way to re-frame the discourse on nuclear weapons by focussing on the humanitarian impact of a nuclear weapon explosion. Through a variety of new and coordinated initiatives, they planned to challenge orthodox security thinking, break through the paralysis that existed in multilateral disarmament fora and create a strong sense of urgency for nuclear disarmament.[47]

Key to this was a loose coalition organised around the International Campaign to Abolish Nuclear Weapons (ICAN),[48] which had been established in Melbourne, Australia, in 2007. This cluster of like-minded anti-nuclear activists became convinced that the restraints of 'the dominant deterrence paradigm' were beginning to weaken and that public interest in nuclear abolition was becoming increasingly politically powerful.[49] Despite the signing of the New START treaty and the US-Russia 'reset', around 2011–12 a decision was taken to focus on humanitarian impacts of nuclear weapons in order to move this agenda forward, and in March 2013 (NATO member) Norway hosted the first Humanitarian Impacts Conference in Oslo. Follow-up conferences were held in Nayarit in Mexico and Vienna, Austria, in 2014, laying the foundations for the 2015 NPT Review Conference. As Alexander Kmentt put it, the idea was to challenge 'The traditional "state-centred" security focus with which nuclear armed States dominate the nuclear weapons debate in existing multilateral fora ... to consider a global human security perspective'.[50]

[46] See Reaching Critical Will, 'NPT action plan monitoring reports, March 2015, https://www.reachingcriticalwill.org/resources/publications-and-research/publications/5456-npt-action-plan-monitoring-reports. The 2010 action plan also reaffirmed the earlier commitment for the negotiation of a Middle East NWFZ.

[47] Alexander Kmentt, *The treaty prohibiting nuclear weapons* (London: Routledge, 2021), p. 19.

[48] For a discussion of this, see Tilman Ruff, 'Negotiating the UN treaty on the prohibition of nuclear weapons and the role of ICAN', *Global Change, Peace & Security* 30:2 (2018), pp. 233–41.

[49] Tom Sauer & Joelien Pretorius, 'Nuclear weapons and the humanitarian approach', *Global Change, Peace & Security* 26:3 (2014), p. 241.

[50] Kmentt, *The treaty prohibiting nuclear weapons*, p. 20.

The five NWS boycotted all three conferences, and made it clear that this was not a viable alternative to the NPT structure.[51] According to those driving the initiative, throughout this time, the objective had been to develop new momentum *inside* the NPT framework, but after the failure of the 2015 Review Conference attention turned to the possibility of a Nuclear Ban Treaty through the UN General Assembly.[52]

Two years later, the TPNW was agreed at the UN. As Article I explained:

> Each Party undertakes never under any circumstances to: (a) Develop, test, produce, manufacture, otherwise acquire, possess or stockpile nuclear weapons or other nuclear explosive devices; (b) Transfer to any recipient whatsoever nuclear weapons or other nuclear explosive devices or control over such weapons or explosive devices directly or indirectly; (c) Receive the transfer of or control over nuclear weapons or other nuclear explosive devices directly or indirectly; (d) Use or threaten to use nuclear weapons or other nuclear explosive devices; (e) Assist, encourage or induce, in any way, anyone to engage in any activity prohibited to a State Party under this Treaty; (f) Seek or receive any assistance, in any way, from anyone to engage in any activity prohibited to a State Party under this Treaty; (g) Allow any stationing, installation or deployment of any nuclear weapons or other nuclear explosive devices in its territory or at any place under its jurisdiction or control.[53]

The treaty also included a provision in Article VI on 'Victim Assistance and Environmental Remediation', this addition being designed to amplify many previously marginalised voices and viewpoints, including those of nuclear test veterans, and representatives of states where nuclear tests had been carried out.

The announcement of the treaty was met with strong condemnation by the five 'recognised nuclear weapons states'. In a rare moment of unity, representatives from the US, Russia, UK, France, and China all made it clear that their respective governments would not sign the treaty, that it didn't recognise their security interests, and that it risked damaging the established

[51] Although several 'nuclear umbrella states' attended, including Australia, Germany, Japan, and Norway.
[52] Kmentt, *The treaty prohibiting nuclear weapons*, p. 104.
[53] UN, 'Treaty on the prohibition of nuclear weapons', Office for Disarmament Affairs, https://disarmament.unoda.org/wmd/nuclear/tpnw/.

architecture centred on the NPT.[54] As the Biden administration's '2022 Nuclear Posture Review' put it:

> The United States does not share the underlying assumption of the TPNW that the elimination of nuclear weapons can be achieved irrespective of the prevailing international security environment. Nor do we consider the TPNW to be an effective tool to resolve the underlying security conflicts that lead states to retain or seek nuclear weapons.[55]

The same was true for India and Pakistan—not members of the NPT, but also united in their rejection of the TPNW. As a statement on the Indian Ministry of External Affairs website explained:

> India does not support the Treaty, and shall not be bound by any of the obligations that may arise from it. India believes that this Treaty does not constitute or contribute to the development of customary international law; nor does it set any new standards or norms.[56]

The treaty was also criticised by a number of prominent scholars. As Michal Onderco at the University of Rotterdam explained in a 2017 article, the TPNW would 'weaken the position of international law, and even if it led to nuclear disarmament, it would put premium on cheating on international commitments'.[57] Others highlighted the fact that it was never likely to achieve actual disarmament given that the nuclear-armed states were never going to sign. However, as director of the US Project on Nuclear Issues, Heather Williams, has pointed out, one interpretation was that the treaty was as much about challenging the narrative around nuclear weapons and an

[54] A different view was taken by Alexander Kmentt: 'At a time when security challenges are on the rise and without credible leadership on nuclear disarmament by States that rely on these weapons, the TPNW provides much needed normative reinforcement of nuclear non-proliferation efforts. In my view, it is the double-standard in the NPT and the lacklustre implementation of its Article VI that weakens it, not the TPNW with its new normative standard to redress exactly those weaknesses.' Kmentt, *The treaty prohibiting nuclear weapons*, p. 214.

[55] '2022 nuclear posture review', US Department of Defense, p. 19, https://media.defense.gov/2022/Oct/27/2003103845/-1/-1/1/2022-NATIONAL-DEFENSE-STRATEGY-NPR-MDR.pdf.

[56] 'India's view on the Treaty on the Prohibition of Nuclear Weapons', Ministry of External Affairs, Government of India, 22 January 2021, https://www.mea.gov.in/press-releases.htm?dtl/33409/Indias+view+on+the+Treaty+on+the+Prohibition+of+Nuclear+Weapons.

[57] Michal Onderco, 'Why nuclear weapon ban treaty is unlikely to fulfil its promise', *Global Affairs*, 3:4–5 (2017), pp. 391–404.

attempt to reframe the discourse than it was about a formulating a concrete and credible pathway to disarmament.[58]

In general, the treaty was well received by the non-nuclear-armed members of the NPT, and broadly seen as supportive of Article VI rather than in opposition. As the former director of the Frankfurt Peace Research Institute, Harald Müller, and Carmen Wunderlich from the University of Duisburg-Essen put it: 'What looks, for the NWS, like an undesirable intrusion onto their turf represents, for ban supporters, an act of self-empowerment in an area they regard as crucial to their own security and survival'.[59] Over the next few years treaty proponents worked tirelessly to secure signatures and ratifications, and in 2021 the TPNW became law (having reached the required 50 signatories). A few voices spoke about the TPNW replacing the NPT—in 2021, Joelien Pretorius from the University of the Western Cape and Tom Sauer from Universiteit Antwerpen wrote that 'The Treaty on the Non-Proliferation of Nuclear Weapons (NPT) has become an impediment to a world free of nuclear weapons, and it is time to move beyond it'.[60] But most TPNW signatories still saw a value in an albeit flawed NPT—as a senior Brazilian diplomat told me in 2023: 'We didn't want the TPNW to interfere with the NPT. And it was definitely not intended to polarise'.[61]

As one of the brightest minds in the UK nuclear community Matthew Harries put it in a response to Pretorius and Sauer, 'The more likely consequence, in my view [of abandoning the NPT], is a world in which disarmament remains just as difficult to achieve but proliferation is now more likely'.[62] Most signatories agreed with this perspective, and the fact that the NPT remained a forum where the NWS could at least be held publicly to account, and proliferation could in theory be managed. While the NPT Review Conference didn't reach consensus at the postponed 2022 meeting, it came surprisingly close to it, and only failed due to a Russian block.[63]

[58] Heather Williams, 'A nuclear babel: Narratives around the Treaty on the Prohibition of Nuclear Weapons', *The Nonproliferation Review* 25:1–2 (2018), pp. 51–63.

[59] Harald Müller & Carmen Wunderlich, 'Nuclear disarmament without the nuclear-weapon states: The nuclear weapon ban treaty', *Daedalus* 149:2 (2020), pp. 171–89.

[60] Joelien Pretorius & Tom Sauer, 'Ditch the NPT', *Survival* 63(4) (2021), pp. 103–24. For similar themes that they explored, see Joelien Pretorius & Tom Sauer, 'When is it legitimate to abandon the NPT? Withdrawal as a political tool to move nuclear disarmament forward', *Contemporary Security Policy* 43:1 (2021), pp. 161–85.

[61] Off-the-record interview in 2023.

[62] Matthew Harries, 'Response: Keep the NPT', *Survival* 63:4 (2021), p. 126.

[63] See Gaukhar Mukhatzhanova, '10th NPT Review Conference: Why it was doomed and how it almost succeeded', *Arms Control Today*, October 2022, https://www.armscontrol.

Whether the regime can weather the storm of great power competition and a more divided global order in the years ahead remains to be seen.[64]

In theory there is no reason why the NPT and the TPNW cannot both exist in relative harmony as the Third Nuclear Age unfolds. And in reality, as the James Martin Centre for Non-proliferation Studies' Jeffrey Knopf explains, the Ban Treaty probably needs a healthy NPT (and maybe even vice-versa):

> Based on the record so far, however, it remains far from clear whether the Ban Treaty can as a practical matter lead to further progress towards nuclear abolition, at least in the near term. If anything, the negotiations seem to have led nuclear-possessing nations to dig their heels in further over the disarmament issue. For this reason, it will be important to supplement the Ban Treaty with continued efforts to restore progress in the traditional approaches to arms control and nonproliferation.'[65]

However, while the Ban Treaty is unlikely to be the path to nuclear disarmament any time soon, it stands as a significant marker and representative totem for the majority of states that are not nuclear armed. Ultimately, it signifies that the loose consensus on the contours of the global nuclear order that were established in the First Nuclear Age, and slowly began to unravel in the Second Nuclear Age, are under major strain in the Third Nuclear Age. Indeed, it is arguably the fact that the nature of the global nuclear order has become so fundamentally challenged that not one but several orders may now be competing. As William Walker put it in a 2025 essay:

> Amidst the debates about disarmament, the tide was already turning against it. The trend gathering momentum after the mid-90s was from convergence to divergence of interests, nuclear ordering to disordering, disarmament to armament, cooperation to conflict, and peace to war, including the active threat of nuclear war. The case for nuclear disarmament became more persuasive just as prospects for its achievement drained away.[66]

org/act/2022-10/features/10th-npt-review-conference-why-it-was-doomed-and-how-it-almost-succeeded. The Review Conference was postponed due to the COVID-19 pandemic.

[64] Rebecca Davis Gibbons & Stephen Herzog, 'Durable institution under fire? The NPT confronts emerging multipolarity', *Contemporary Security Policy* 43:1 (2021), pp. 50–79.

[65] Jeffrey J.W. Knopf, 'Not by NPT alone: The future of the global nuclear order', *Contemporary Security Policy* 43:1 (2021), p. 204.

[66] William Walker, 'Reflections on complexity, nuclear ordering and disordering over time', *Cambridge Review of International Affairs* 38:3 (2025), pp. 280–301.

Despite a growing narrative of crisis, the NPT remains an essential forum for many governments and is still largely seen as the foundation of global nuclear order. Albeit as Bill Potter and Sarah Bidgood put it in a 2025 essay, 'As with Beckett's characters, NPT delegates appear content to pass their time tethered to routines and practices they cannot abandon and conversations they cannot resolve'.[67]

Nuclear culture wars?

According to a 2023 article in the British newspaper *The Guardian*, the Department of Peace Studies at the University of Bradford beat the Department of War Studies at Kings College London 15-0 in a football match in the mid-1990s.[68] The annual match would become known as 'The Tolstoy Cup', and was only cancelled due to the COVID-19 pandemic that began in early 2020: as far as I am aware at the time of writing the match hasn't been replayed. While the ivory tower rivalry between Peace Studies and War/Strategic Studies goes back to the 1970s, the larger contestation between two quite different visions of how to tackle the nuclear problem can be traced back to at least the 1940s.[69] As Chapter 1 explains, nuclear history is dominated by narratives that focus on the most powerful state-based players and concerns about managing nuclear threats. But a parallel and perhaps less pervasive history of protest and contestation against this state of affairs also exists—although it is perhaps not as well known. Of course, challenging dominant nuclear narratives and frameworks is far from new: many disarmament groups were formed, and prominent activists cut their teeth in the early First Nuclear Age and became particularly influential in the 1980s, while an alliance of NGOs, civil society organisations, and scholarship challenging traditional nuclear paradigms flourished and spread in the Second Nuclear Age. But this 'battle of ideas' seems to be becoming more conspicuous as the intellectual and normative environment has shifted and splintered in the Third Nuclear Age. Worryingly there seems to be an

[67] In Bill Potter & Sarah Bidgood, 'Waiting for Godot at the 2026 NPT review conference', *Arms Control Today*, September 2025, https://www.armscontrol.org/act/2025-09/features/waiting-godot-2026-npt-review-conference.

[68] Mark Brown, '"They thought we were all communists": Bradford's peace studies department turns 50', *The Guardian*, 8 April 2023, https://www.theguardian.com/education/2023/apr/08/bradford-university-peace-studies-department-turns-50.

[69] The Department of War Studies opened in 1961 and the Peace Studies Department opened in 1972.

increased tendency for those with different viewpoints to talk past each other rather than to engage. Thus, what might be thought of as the impact of 'nuclear counterculture' needs to be unpacked.

On 9 July 1955, the instantly recognisable scientist Albert Einstein and the prominent British philosopher Bertrand Russell issued a dire warning in their 'manifesto' highlighting the existential dangers of nuclear weapons. Perhaps the most memorable part was the following:

> There lies before us, if we choose, continual progress in happiness, knowledge, and wisdom. Shall we, instead, choose death, because we cannot forget our quarrels? We appeal as human beings to human beings: Remember your humanity, and forget the rest. If you can do so, the way lies open to a new Paradise; if you cannot, there lies before you the risk of universal death.[70]

Nine years earlier Einstein (along with the physicist Leo Szilard) had been influential in establishing the Emergency Committee of Atomic Scientists, perhaps the first major nuclear civil society group. This would be followed by organisations such as Pugwash Conferences on Science and World Affairs (1956), the National Committee for a SANE Nuclear Policy(1957)—later renamed Peace Action, the Campaign for Nuclear Disarmament (CND) (1957), and Women Strike for Peace (1961). Prominent figures of the time, including Dr Martin Luther King, were also strong and vocal supporters of nuclear disarmament.[71]

The growth in advocacy groups against the bomb was mirrored by the publication of several influential books by journalists and members of the academy. In 1946, American writer John Hersey published his horrifying and highly influential account of the impact of the first atomic bomb dropped on Hiroshima.[72] As the arms race intensified, the 1950s and 1960s would see a handful of influential academics and other prominent figures calling for a different future for humanity: Bertrand Russell's, *Common Sense and Nuclear Warfare* (1959)[73] and various works by the

[70] Available at: https://ahf.nuclearmuseum.org/ahf/key-documents/russell-einstein-manifesto/.

[71] See, for example, Vincent J. Intondi, *African Americans against the bomb: Nuclear weapons, colonialism, and the Black Freedom Movement* (Stanford, CA: Stanford University Press, 2015).

[72] The original article actually appeared in *The New Yorker*: John Hersey, 'Hiroshima', *The New Yorker*, 31 August 1946, pp. 15–68.

[73] Bertrand Russell, *Common sense and nuclear warfare* (Hemel Hempsted: George Allen & Unwin Ltd, 1959).

French philosopher John Paul Sartre,[74] followed by American psychologist Charles Osgood's *An Alternative to War or Surrender* (1962),[75] the American mathematician Anatol Rapaport's, *Strategy and Conscience* (1969),[76] and the American political theorist Philip Green's *Deadly Logic: The Theory of Nuclear Deterrence* (1966).[77] In 1968, Green also wrote a prominent critique of the role that think tanks had played in institutionalising a particular view of nuclear weapons.[78]

This protest may well have been significant in helping to pressuring (particularly the US and UK) governments to agree to the PTBT in 1963. But dominance of rationalist, realist thought and the apparent logic of deterrence meant that such visions for an alternative nuclear future remained on the periphery. Indeed, as the treatment of J. Robert Oppenheimer by the US Atomic Energy Agency in the mid-1950s showed, there was little room for dissent inside the nuclear hierarchy.

Disarmament and peace activism would return to prominence in the 1980s, especially in Europe and North America. In the UK the British academic E.P. Thompson published the *Protest and Survive* pamphlet (as a parody of the UK government's issued *Protect and Survive* guidance), and the women's peace camp began at Royal Air Force (RAF) Greenham Common in 1981. Across the Atlantic, fears of a 'new cold war' had driven the founding of the Nuclear Freeze Movement by the Committee for a Sane Nuclear Policy and others, which culminated in a huge rally attended by over one million people in New York in 1982. While in Japan the Hiroshima and Nagasaki Peace Ceremonies (which had existed since the late 1940s) became focal points for international disarmament solidarity.

Anti-nuclear thinking and activism was bolstered by the development of 'Peace Studies' as an academic discipline, rooted in the field of liberalism, which began to present a more prominent challenge to the dominance of realist Strategic Studies towards the end of the First Nuclear Age. Peace

[74] Ben Zala, 'No one around to shut the dead eyes of the human race': Sartre, Aron, and the limits of existentialism in the Nuclear Age', *Review of International Studies* 49:5 (2023), pp. 795–812.

[75] Charles Osgood, *An alternative to war or surrender* (Champaign, IL: University of Illinois Press, 1962).

[76] Anatol Rapoport, *Strategy and conscience* (Tel Aviv: Schocken Books, 1969).

[77] Philip Green, *Deadly logic: The theory of nuclear deterrence* (Columbus, OH: Ohio State University Press, 1966).

[78] Philip Green, 'Science, government, and the case of RAND: A singular pluralism', *World Politics* 20:2 (1968), pp. 301–26.

Studies grew from the work of Immanuel Kant,[79] Johan Galtung,[80] and others, and was driven by a belief that institutions, rules, and norms could manage the nuclear condition, and that disarmament was possible. This scholarship often overlapped with prominent peace and anti-nuclear activist groups such as Greenpeace (founded in 1971), CND, and the International Physicians for the Prevention of Nuclear War (founded in 1980) and became particularly significant in the early 1980s. Notably, such thought impacted UK politics at the time, and the decision by the Labour Party to make unilateral nuclear disarmament a manifesto pledge in 1983. Perhaps indicative of nuclear orthodoxy in the period often referred to as the 'second cold war', Labour lost the election, and the manifesto would later memorably be termed by Labour member of parliament (MP) Gerald Kaufman as the 'longest suicide note in history'.[81] In a similar vein, two years earlier my home city of Leicester was declared a 'nuclear-free zone' in 1981—the extent that this impacted Soviet Cold War targeting strategy remains unknown ...[82]

The result was that while a different vision of humanities' nuclear future emerged during the First Nuclear Age, and challenges to the logic of a system based around deterrence and arms control existed, the intellectual space was dominated by a particular view of what was and wasn't possible. Ivory tower voices and public intellectuals (however prominent) and disarmament-focussed NGOs remained largely peripheral players.

As the world moved into a Second Nuclear Age in the early 1990s, two significant developments combined. The first was a shift in the academy away from the dominance of traditional Strategic Studies, especially in Europe, towards more 'critical' scholarship. The second was in effect the globalisation of anti-nuclearism as a number of new NGOs were formed, civil society rallied, and links were built with the more activist disarmament-focussed states. The broadly more peaceful and stable environment of the Second Nuclear Age created the space for challenges to the nuclear mainstream to flourish, and the intellectual space within which nuclear politics was thought about became more contested—at least in the Western world.

[79] Immanuel Kant, *Perpetual peace: A philosophical sketch* (1795).

[80] For example, Johan Galtung, 'Alternatives to the nuclear arms race: Ten proposals for concrete peace politics for the 1980s', *Bulletin of Peace Proposals* 12:4 (1981), pp. 361–4.

[81] Rajini Vaidyanathan, 'Michael Foot: What did the "longest suicide note" say?', *BBC News Magazine*, 4 March 2010, http://news.bbc.co.uk/1/hi/magazine/8550425.stm.

[82] Many local authorities in the UK did likewise, and joined the Nuclear Free Local Authorities Network, https://www.nuclearpolicy.info.

In North America, the dominance of traditional realist Strategic Studies was increasingly being challenged by liberal and even constructivist scholarship. While Carol Cohn's seminal article 'Sex and Death in the Rational world of Defense Intellectuals' marked a key moment in the development of feminist critiques of nuclear orthodoxy,[83] perhaps the best example was the publication of Nina Tannenwald's thought-provoking article on 'The Nuclear Taboo' in 1999.[84] The shift was significantly pronounced enough for American political scientist Richard K. Betts to plead in 1997 for Strategic Studies to retain 'its status equal to any other subfield' as scholarship mirrored what many people believed to be a fundamental transformation in international politics at the end of the Cold War.[85] But what the then Aberystwyth-University-based academics Campbell Craig and Jan Ruzicka described as a powerful 'non-proliferation complex' continued to exert significant influence across the US academy, and work within the Strategic Studies lexicon remained dominant in many leading universities and think tanks, both of which continued to receive substantial funding, albeit that this began to change in 2021, after a shift in funding priorities of major donors.[86]

The story in Europe and in other parts of the 'Western world' is slightly different. Outside the US, the move into the Second Nuclear Age coincided with gradual decline in scholarship that might be labelled as traditional Strategic Studies. As in the US, liberalism enjoyed a boost, but the academic zeitgeist appeared to be moving further away from the Cold War canon. Increasingly scholars working on nuclear issues reflected the broader shift in the social sciences towards post-positivist, constructivist, and critical approaches to the study of security. This can probably be traced back to Aberystwyth University and to a conference held at York University in 1994,[87] and was fully fleshed out by the prominent academics Richard Wyn Jones in *Security, Strategy and Critical Theory* in 1999,[88] and by Barry

[83] Carol Cohn, 'Sex and death in the rational world of defense intellectuals', *Signs: Journal of Women in Culture and Society* 12:4 (1987), pp. 687–718.

[84] Nina Tannenwald, 'The nuclear taboo: The United States and the normative basis of nuclear non-use', *International Organization* 53:3 (1999), pp. 433–68.

[85] Richard K. Betts, 'Should strategic studies survive?' *World Politics* 50:1 (1997), p. 33.

[86] E.g. Brian Bender, 'A big blow: Washington's arms controllers brace for loss of biggest backer', *Politico*, 19 July 2021, https://www.politico.com/news/2021/07/19/washington-arms-controllers-nuclear-weapons-500126.

[87] And, specifically, to a conference held at York University in 1994. See Keith Krause & Michael C. Williams (eds.), *Critical security studies: Concepts and cases* (Minneapolis, MN: University of Minnesota Press, 1997).

[88] Richard Wyn Jones, *Security, strategy and critical theory* (Boulder, CO: Lynne Rienner, 1999).

Buzan, Ole Waever, and Jaap de Wilde in their 1998 book *Security: A New Framework for Analysis*. As Buzan et al explained:

> Two schools of thought now exist in security studies: traditionalists want to restrict the subject to politico-military issues; while wideners want to extend it to the economic, societal and environmental sectors.[89]

In particular, this movement drew attention away from the focus on 'nation states' and interstate war as the main referent for Security Studies and towards issues such as 'human security'—arguably the forebear for the Humanitarian Initiative 15 years later. The distinguished British scholar Ken Booth, captured this transition in his book, *Theory of World Security* in 2007.[90]

As a result, the Second Nuclear Age would see an expansion of scholarship looking beyond 'traditional hard security and military strategy', increasingly utilising 'constructivist' and 'critical' approaches such as post-modernism, feminism, post-structuralism, and post-colonialism. This trend was also reflected in the broader academic milieu as non-Western scholars writing about nuclear politics such as Itty Abraham, Mohammad Ayoob, Shampa Biswas, Abena Dove Osseo-Asare, and others lent their perspectives to this burgeoning body of 'critical thought'.[91]

By the early 2010s, this manifested in the birth of Critical *Nuclear* Studies. Critical Nuclear Studies sought to challenge traditional security paradigms by emphasizing human security, ethics, discourse, and power structures. It sought to provide alternative perspectives on nuclear policy, advocating for disarmament, decolonization, and a more inclusive approach to security. As such, Critical Nuclear Studies was designed to build on disarmament activism and Peace Studies in seeking to challenge established practices of nuclear deterrence, and at the same time, draw on progressive

[89] Barry Buzan, Ole Waever, & Jaap de Wilde, *Security: A new framework for analysis* (Boulder, CO: Lynne Rienner, 1998).

[90] Ken Booth, *Theory of world security* (Cambridge: Cambridge University Press, 2007).

[91] E.g. Itty Abraham, *The making of the Indian atomic bomb: Science, secrecy and the post-colonial state* (London: Zed Books, 1998); Shampa Biswas, *Nuclear desire: Power and the postcolonial nuclear order* (Minneapolis, MN: University of Minnesota Press, 2014); Abena Dove Osseo-Asare, *Atomic junction: Nuclear power in Africa after independence* (Cambridge: Cambridge University Press, 2019).

Mohammed Ayoob is also important to mention here albeit his most influential work, 'Security in the third world: The world about to turn', was published in *International Affairs* in 1983.

and emancipatory International Relations (IR) theory to amplify different voices and opinions within the global nuclear order.

This new subfield of nuclear politics was probably first captured by Australian IR scholar Anthony Burke in a special issue of the journal *Critical Security Studies* titled: 'Nuclear Politics beyond Positivism'.[92] As Burke explained in the introduction:

> What is it possible to say, or think, about nuclear weapons today? The answer will depend less on what you know than on who you are: on where you are situated, and on what the background rules of memory, discourse and research that shape your community's understanding of the nuclear problem are.[93]

Critical nuclear scholarship also spoke directly to those negatively impacted by nuclear weapons in the past,[94] or that might be in the future. By the mid-2010s, 'critical' approaches to nuclear issues were becoming the dominant approach in academic scholarship in the UK, and influential—albeit to a slightly lesser extent—in Australia and Europe, but also increasingly in institutions across the world. Work—especially that by the erudite British scholar Nick Ritchie—began to appear on devaluing or stigmatizing nuclear weapons and deterrence,[95] and providing different perspectives on the nuclear condition.

At the same time, the range and influence of NGOs expanded considerably during this period.[96] In 1995 Joseph Rotblat and Pugwash were awarded the Nobel Peace Prize 'for their efforts to diminish the part played by nuclear arms in international politics'.[97] The same year, the Australian government launched 'The Canberra Commission' to find practical steps towards

[92] Anthony Burke, 'Nuclear politics: Beyond positivism', *Critical Studies on Security* 4:1 (2016), p. 1.

[93] Burke, 'Nuclear politics', p. 1.

[94] E.g. Traci Brynne Voyles, *Wastelanding: Legacies of uranium mining in Navajo country* (Minneapolis, MN: University of Minnesota Press; 2015); and Becky Alexis Martin, *Disarming doomsday: The human impact of nuclear weapons* (London: Pluto Press, 2019).

[95] See Nick Ritchie, 'Valuing and devaluing nuclear weapons', *Contemporary Security Policy* 34:1 (2013), pp. 146–73; and Nick Ritchie, 'Waiting for Kant: Devaluing and delegitimizing nuclear weapons', *International Affairs* 90:3 (2014), pp. 601–23.

[96] For a discussion of this and the impact, see Michal Onderco, 'Civil society and nuclear risk reduction', Wilson Center, 10 March 2025, https://www.wilsoncenter.org/publication/civil-society-and-nuclear-risk-reduction.

[97] The Nobel Prize, 'Summary', The Nobel Peace Prize 1995, NobelPrize.org, Nobel Prize outreach 2026, Friday, 13 March 2026, https://www.nobelprize.org/prizes/peace/1995/summary/.

nuclear disarmament.[98] In 1999, Reaching Critical Will was founded, and a year later Abolition 2000. But as the decade now known as the 'noughties' unfolded, this momentum picked up: in 2007 ICAN was launched in Melbourne, Australia; ICAN would of course be central to the agreement of the Nuclear Ban Treaty a decade later. In 2008, Global Zero was formed, and unveiled the 'global zero action plan' outlining an ambitious 10-year pathway to disarmament.[99] While the influence of these group perhaps began to wane in the early 2020s, in 2024, the Nobel Peace Prize was awarded to Nihon Hidankyo, an organisation of survivors from the 1945 atomic bombings of Hiroshima and Nagasaki in recognition of their tireless work to educate about nuclear risks and disarmament.

The rise of 'Critical Nuclear Studies' and what former UK Ministry of Defence official and UN disarmament commissioner, Paul Schulte, has described as an 'archipelago' of critical security scholars, NGOs, and activist disarmament groups,[100] appeared to be gradually challenging established, positivist dominance of the nuclear debate outside of government. Importantly, they increasingly had buy-in from officials and politicians from a growing range of non-nuclear-armed states. This was significant because it would shape how policy elites in different countries view nuclear weapons, the way that wannabe future thought leaders and officials studying political science and IR view the world, and the types of nuclear futures that one might imagine.

Of course, as an academic who spends many hours reading about such things, I think this intellectual story is important, but why is it important to how we understanding the politics of the Third Nuclear Age? The reason is that this academic context, civil society influence, and intellectual ecosystem directly and indirectly shapes how policy elites think about the nuclear condition, the legitimate scope of what constitutes 'nuclear knowledge', and arguably what might be possible in the years ahead.

[98] The Canberra Commission on the Elimination of Nuclear Weapons, 'Statement', https://www.ccnr.org/canberra.html.

[99] Global Zero, 'Action plan', https://www.globalzero.org/reaching-zero/index.html.

[100] Paul Shulte, 'Conflicting views of a darkening strategic prospect: The opposed sensibilities and discordant perspectives of the nuclear practitioners and disarmament archipelago', in Brad Roberts (ed.), *Morality and nuclear weapons: Practitioner perspectives* (Livermore, CA, and Grenoble: Center for Global Security Research, Lawrence Livermore National Laboratory in partnership with the French Commission on Atomic and Alternative Energy, July 2023), https://cgsr.llnl.gov/sites/cgsr/files/2024-08/CGSROccasional_Paper_MoralityandNuclearWeapons_06302023.pdf.

A key reason for this is that nuclear scholarship has always played a key role in informing policy. For most of the period since 1945, nuclear studies research has been focussed on solving nuclear policy 'problems' through mechanisms such as deterrence, strategic stability, and arms control. The main 'customer' for this research were nuclear-armed governments, especially the US. However, today this may have changed: the main customers for critical nuclear research are NGOs, civil society groups, and increasingly diplomats and officials from non-nuclear-armed states. In some ways this resembles the nexus between Peace Studies scholars and disarmament activists in the past, but the impact today seems much more tangible and global.

For example, critical scholars have worked alongside ICAN since it was established in 2007,[101] and as the University of Leeds' Laura Considine explains, the development of critical nuclear literature played an important role in the Humanitarian Impacts Initiative.[102] Critical nuclear scholarship has also played a role in amplifying the voices of previously marginalised communities within the global nuclear order, particularly those where nuclear testing or other exploitation occurred.

But whether this can continue into the future is a moot point. It is possible that the politics of the Third Nuclear Age will mean less influence for NGOs and non-state opinion formers who were so temporarily influential in bringing about the diplomatic exploit of the TPNW. As Paul Schulte put it,

> there are reasons to expect the Archipelago and civilian activism, to diminish in international importance and leverage over nuclear decision-makers, with a number of implications for the nuclear future.[103]

It may even be possible that we see a ramping up of efforts by current nuclear-armed states and their governments to bolster support for deterrence. Indeed, in 2023 the UK government announced that it would be establishing a 'deterrence fund'.[104] In some ways harking back to the

[101] As documented in Kmentt's *The treaty prohibiting nuclear weapons*, pp. 18–19.

[102] Laura Considine, 'The "standardization of catastrophe": Nuclear disarmament, the Humanitarian Initiative and the politics of the unthinkable', *European Journal of International Relations* 23:3 (2016), pp. 681–702.

[103] Email exchange with Paul Schulte.

[104] George Allison, 'Britain launches £3.3m "nuclear deterrence fund"', *UK Defence Journal*, 18 August 2024, https://ukdefencejournal.org.uk/britain-launches-3-3-million-nuclear-deterrence-fund/.

academic positions that both the US and the UK previously supported during the Cold War.

It is probably fair to say that as of 2025 the balance between scholars working from a 'traditional' Strategic Studies background and those working from a critical approach is roughly in equilibrium globally. This is a far cry from the academic landscape of the First and Second Nuclear Ages. As my friend and colleague Ben Zala put it 'if you were at an academic conference focussing on security studies in the First Nuclear Age "you probably didn't need to ask someone what they worked on"'. If you asked that question today the answer would likely be quite different. The result is that as the Third Nuclear Age unfolds, and perhaps as former colonised states in the non-nuclear world build up their centres of knowledge production, the academic and intellectual environment within which nuclear politics is thought about, and which indirectly shapes the contours of the policy context, will be far less coherent than at any time before. To put it simply, traditional Strategic Studies and realist and liberal paradigms that focussed on *managing the nuclear condition* that dominated the past are being fundamentally challenged by different ways to *transform the nuclear condition* by critical scholars. In theory this should make for a more democratised intellectual environment within which nuclear issues can be discussed, but in reality it may make debate and discussion more difficult.

Despite this trend, there is still a feeling that certain types of scholarship play a key role in limiting what is possible. As the brilliant Benoît Pelopidas at Science Po put it:

> In the current configuration, we think of ourselves, at best, as responsible *managers* of the nuclear *present*, a present expected to be extended into the future. If we become aware of our self-imposed blinders, we could conceive ourselves as responsible *citizens* of a *future* in the making.[105]

By way of an example both traditional security studies scholars and critical scholars appear to feel that the other has more backing, voice, and impact.[106]

[105] Benoît Pelopidas, 'Nuclear weapons scholarship as a case of self-censorship in Security Studies', *Journal of Global Security Studies* 1:4 (November 2016), pp. 326–336.

[106] As Kmentt explains: 'The perspective of non-nuclear weapon States only plays a marginal role in the work of most representatives of the strategic studies community. The small number of researchers who do write with this perspective in mind, cannot compensate for the clear dominance of the orthodox, and in comparison, well-funded, nuclear weapons approach that academia and think tanks are accustomed to follow ... The combined forces of the States

A good illustration of this is provided by Alexander Kmentt, as he argues in his 2021 book:

> Finally, academia and think tanks within the strategic studies community, i.e. those that deal with security policy issues and in particular nuclear weapons, also play a significant role in shaping the nuclear weapons discourse and the perception of the TPNW. The vast majority of actors in the strategic studies community are concentrated in nuclear armed States or nuclear umbrella States. In those States, a certain 'nuclear weapons culture' of shared practices and attitudes has developed over a long period of time, and most tend to look at security, nuclear weapons and nuclear deterrence in a certain way. This particular 'lens' is overwhelmingly in line with the traditional parameters of the nuclear weapons discourse, as determined by the States that dominate the nuclear status-quo, and is an important reinforcing factor.[107]

The concern is that this leads us to a position whereby 'each side' cannot—or is unwilling to—understand the logic of the 'opponent', and often different views are demonised. This may reflect an increasing cultural trend to deal in absolutes rather than dialogue and debate, a dynamic that continues to leak into the academic and professional nuclear space. This might not just mean disagreement, but an increasing inability or unwillingness to engage with different viewpoints—a sort of 'nuclear siloing'. Or as the Nuclear Threat Initiative's Doug Shaw put it to me several years ago, 'ignoring voices rather than defeating arguments'.[108]

The *nuclear* culture wars between governments and activists, traditional strategic studies, and critical scholars are far less prominent that the battles that have raged over recent years across Western societies about social norms, morality, and identity, but the clash between quite different visions of the nuclear future is important. This is particularly the case given the central role that universities, think tanks, research institutes, NGOs, (and the media) play in educating the next generation of thought leaders when it comes to nuclear issues. Just as realism and strategic studies may have inculcated a certain view of the nuclear world in the First Nuclear Age, and

opposing the Humanitarian Initiative and the TPNW together with most of the orthodox strategic studies community make for formidable opponents for the supporters of the TPNW.' In Kmentt, *The treaty prohibiting nuclear weapons*, p. 224.

[107] Kmentt, *The treaty prohibiting nuclear weapons*, p. 223.

[108] Email correspondence with the author.

the promises of liberalism emerged in the Second Nuclear Age, it is possible that critical scholarship will open up new avenues for contestation and perhaps transformation in the Third Nuclear Age. This will become particularly pronounced as former 'developing' or 'global south' countries build up their centres of knowledge production and challenge the overwhelming dominance of institutions and learned publications based in the West. While possibly far-fetched, in the longer term winning the nuclear culture war could become as important as deploying the latest military technologies.[109]

Conclusion: Alternative nuclear futures ...

It is easy to look back on nuclear history with the benefit of uncritical hindsight and suggest that a certain consensus and an accepted 'order' existed across the First Nuclear Age and the first half of the Second Nuclear Age, but of course this masks significant disagreement, divergence, and different visions for that particular order. Nevertheless, three things are happening in the Third Nuclear Age that may well by undermining the foundations of this nuclear orthodoxy: the erosion of agreed frameworks of nuclear arms control and the shared belief that restraint and risk reduction are mutually beneficial; a challenge to the legitimacy and fairness of the governance and management structures set up over previous decades to address nuclear issues; and a shift in the intellectual ecosystem within which nuclear politics is thought about and how nuclear knowledge is created, legitimised, and contested.

What this means for nuclear politics in the Third Nuclear Age isn't clear cut. It is possible that Washington and Moscow will find a reason to return to arms control, but this is unlikely to be the basis for sustained cuts and other risk reduction measures that characterised the past. Beyond this dyad, the picture doesn't get much more reassuring. Arms control involving India and Pakistan is not just spoiled by historical animosity and distrust, it is also complicated by the trilateral relationship involving China. China in turn is unlikely to want to 'lock-in' what may be perceived as historical inequalities; the same is true *vis-à-vis* the UK and France as they confront a new European security complex. While arms control involving Israel or North Korea is also difficult to visualise at the current time. It is not inconceivable that additional states will leave the NPT, damaging the non-proliferation norm that

[109] I am grateful to Shivani Singh for this thought-provoking point.

has stood at the centre of international politics for over 50 years. This may mean that non-nuclear-armed states see other mechanisms as more conducive to meeting their security and disarmament desires. Whether this will represent a better way to manage nuclear dangers is a moot point. Lastly, it isn't clear who or what will 'win' the battle of ideas that undergirds how society thinks about nuclear technologies, or indeed what this victory will look like. Nevertheless, it is hard to escape the logic that each of these dynamics will lead to a further entrenchment of the belief in the value of nuclear weapons for a small group of states rather than the opposite.

To paraphrase the American academic Nina Tannenwald it seems that at least two quite different—and broadly antagonistic—nuclear orders have come into existence:[110] one involving the vast majority of states, committed to both nuclear non-proliferation and disarmament, and wanting to break out of a condition of nuclear captivity; and a second where nuclear armament means power, and deterrence is a fundamental component of security in a competitive and dangerous world. The clash between these two visions will define the politics of the Third Nuclear Age.

[110] Nina Tannenwald, 'The great unravelling: The future of the nuclear normative order', in Nina Tannenwald and James M. Acton (eds.), *Meeting the challenges of the new nuclear age: Emerging risks and declining norms in the age of technological innovation and changing nuclear doctrines* (Cambridge, MA: American Academy of Arts & Sciences (2018), p. 6, https://www.amacad.org/sites/default/files/publication/downloads/New-Nuclear-Age_Emerging-Risks.pdf.

6
Atoms and the Anthropocene
The nuclear-climate-development nexus

Introduction: Putting the nuclear challenge into context

Nuclear technologies have always exhibited characteristics similar to Robert Louis Stevenson's 1886 Gothic horror novel, *The Strange Case of Dr Jekyll and Mr Hyde*. On one hand, like the kind and respected Dr Jekyll, the potential application of nuclear technologies to produce massive amounts of 'clean' energy, open up pathways to sustainable economic development, drive forward breakthroughs in medicine, and transform our understanding of the earth, outer space, and science in general are tantalising glimpses of a utopian future. On the other hand, like the evil and destructive Mr Hyde, nuclear technologies have been intrinsically infused with the dystopian reality that they can kill indiscriminately, and destroy not just military forces or a huge number of people in cities, but potentially our whole planet—deliberately, or even by accident. This is a paradox that Columba Peoples at the University of Bristol has termed 'Nutopia': 'a mode of understanding nuclear power that is imbued with a spirit of technological optimism in relation to "peaceful" nuclear power, but simultaneously qualified by an awareness of the destructive uses and catastrophic potentialities of nuclear weapons.'[1]

This difficult balance has always been part of the nuclear condition, and was perhaps best summed up by United States (US) President Dwight Eisenhower in a 1953 speech to the United Nations (UN) General Assembly:

> the United States pledges to you—and therefore before the world—its determination to help solve the fearful atomic dilemma—to devote its

[1] Columba Peoples, 'Redemption and Nutopia: The scope of nuclear critique in international studies', *Millennium* 44:2 (2016), p. 1.

The Third Nuclear Age. Andrew Futter, Oxford University Press. © Andrew Futter (2026).
DOI: 10.1093/9780198944126.003.0007

> entire heart and mind to find the way by which the miraculous inventiveness of man shall not be dedicated to his death, but consecrated to his life.[2]

The address would come to be known as the 'Atoms for Peace' speech and the 'right' of all nations to access civilian nuclear technology would become enshrined in Article IV of the Non-Proliferation Treaty (NPT) 15 years later.[3] But in the ensuing decades, nuclear technologies have not been spread evenly or equally, with most of 'benefits' being 'enjoyed' by those with the most advanced economies in the developed world, and the locus of dangers being centred on a small minority with nuclear arms, nuclear research institutes, and a slightly larger group that operate nuclear powerplants.

In the Third Nuclear Age we are again revisiting the Nutopian conundrum posed by nuclear technologies as humanmade system-level change is now widely recognised to be altering our planet. This has been driven in part by the perceived role of the atom to drive forward economic and societal development and at the same time form part of the solution to the now widely accepted existential challenge of environmental climate change. This has created a nuclear-climate-development trifecta that is integrated and co-constitutive: the spread of nuclear technologies raises the possibility that nuclear use or accidents will impact the climate and humanity; exponentially increasingly climate, scientific, and developmental demands all seem to require more—not less—nuclear technology; and for the first time ever, climate change and developmental dynamics could impact nuclear security, safety, and stability. The use of nuclear technology remains contentious for a variety of reasons, and old challenges of safety, security, and waste storage endure, but it is hard to escape the fact that the enormous possibilities of spitting the atom are increasingly attractive to many different actors. Taken together these dynamics are driving us further into what some have started to call 'the Anthropocene'—a new era of planetary geological time shaped by humans.

[2] 'Text of the address delivered by the president of the United States before the General Assembly of the United Nations in New York City Tuesday afternoon, December 8, 1953', Dwight D. Eisenhower Presidential Library, https://www.eisenhowerlibrary.gov/sites/default/files/file/atoms_Binder13.pdf.

[3] For a deeper discussion of Article IV and the right to nuclear technologies, see Chapter 4.

Nuclear winter and planetary destruction

In an interview shortly after the first nuclear test that took place in the Alamogordo Desert in New Mexico in July 1945, lead scientist on the Manhattan Project, J. Robert Oppenheimer uttered the phrase from the Bhagavad Gita: 'now I am become death; the destroyer of worlds'.[4] While the exact meaning of those words in the ancient Hindu script is debated, Oppenheimer wasn't overstating things. Never before had humans built something so immediately and immensely destructive—the yield of the 'trinity test' was estimated at 21 kilotons—or the equivalent of 21,000 tons of conventional explosives. The blast also propelled significant amounts of Iodine 131, Caesium 137, Strontium 90, and other nasty radioactive by-products into the desert air, impacting many local indigenous people and their livelihoods. Thus began the connection between nuclear technologies and their negative impact on our environment, climate, and planet.

For the best part of the next 20 years, until the signing of the Limited Test Ban Treaty in 1963, first the US, then the Soviet Union, and later Britain and France conducted hundreds of explosive nuclear tests, most of them above ground, in the atmosphere, or at sea, releasing enormous amounts of radiation and causing significant human and environmental damage.[5] Most of these subsequent tests were far larger than the trinity explosion, or the 15 kiloton 'Little Boy' bomb dropped on Hiroshima, with yields soon measured in megatons (millions of tons of TNT) rather than thousands of tons of TNT.[6] By way of comparison, the fire-bombing of the German city of Dresden in February 1945 by the British Royal Air Force (RAF) and US army air force caused similar structural damage to that in Hiroshima, but involved hundreds of aircraft dropping thousands of bombs over two days (killing around 130,000 people).[7] The explosive yield of a single megaton weapon is probably on a scale similar to *all* of the weapons used by *all* of the combatants during World War Two.

[4] The full quote is: 'We knew the world would not be the same. A few people laughed, a few people cried, most people were silent. I remembered the line from the Hindu scripture, the Bhagavad-Gita. Vishnu is trying to persuade the Prince that he should do his duty and to impress him takes on his multi-armed form and says, "Now, I am become Death, the destroyer of worlds". I suppose we all thought that one way or another.' Available from The Atomic Archive, https://www.atomicarchive.com/media/videos/oppenheimer.html.

[5] See, for example, Richard Miller, *Under the cloud: The decades of nuclear testing* (Texas: Two Sixty Press, 1991).

[6] 1 ton of TNT is probably enough to destroy a small building.

[7] See Sinclair McKay, *The fire and the darkness: The bombing of Dresden 1945* (St Martin's Press, 2020).

By the 1950s, and particularly after the US Castle Bravo test in 1954 that led to the death of several Japanese fishermen aboard the sadly-ironically named boat 'Lucky Dragon',[8] environmental groups and the public were becoming increasingly aware of the effects of nuclear testing, but particularly what it might mean if these weapons were ever used in war. In 1957, British-Australian novelist Neville Shute published the book *On the Beach* depicting the slow extinction of humanity as a result of radiation spreading across the planet following a nuclear war in the northern hemisphere.[9] Two years later, American author Pat Frank published *Alas Babylon* depicting societal breakdown in a post-nuclear US town called Fort Repose.[10] As if to confirm these fears, in October 1961, the Soviet Union tested the 'Tsar Bomba' high above Severny Island in the Arctic Circle. The test was later estimated to have a yield of 50 megatons—nearly 2,000 times more powerful than the bombs dropped on Hiroshima and Nagasaki (the bomb had a three-minute fuse, which was just enough time for the pilots to get away). A year later, the US and the Soviet Union came as close to nuclear war as at any other time since, in what became known as the Cuban Missile Crisis. It was becoming increasingly apparent that if war happened, enough nuclear weapons had already been built to completely destroy the planet and all those who inhabited it.[11]

Two decades later, as the nuclear rivalry between East and West appeared to reach a new nadir and what some began to describe as a 'new cold war',[12] eminent US scientist Carl Sagan and others began warning about the possibility of a 'nuclear winter' caused by the fallout from even a so-called 'limited' nuclear exchange. Essentially, and in addition to the destruction and death caused the nuclear blast, heat wave, radiation, and fallout, the masses of smoke and debris caused by nuclear explosions would block out the sun, causing a prolonged period of darkness that would lead to a significant drop in global temperatures, crop failures, and possibly mass starvation.[13] The nuclear winter thesis would be revisited in the 2010s, with some scholars arguing for 'winter-safe deterrence' whereby certain types of nuclear use—and a small stockpile of warheads—may not cause catastrophic

[8] Oishi Matashichi, *The day the sun rose in the west: Bikini, the Lucky Dragon, and I* (Honolulu: University of Hawai'i Press, 2011).
[9] Nevil Shute, *On the beach* (New York: William Morrow and Company, 1957).
[10] Pat Frank, *Alas, Babylon* (Rockville, MD: Wildside Press LLC, 1959).
[11] By 1963, there were over 30,000 nuclear weapons in existence, the overwhelming majority being in the US nuclear stockpile. See 'Global nuclear stockpiles, 1945–2006', *Bulletin of the Atomic Scientists* 62:4 (2004), pp. 64–6.
[12] E.g. Fred Halliday, *The making of the second cold war* (London: Verso Books, 1983).
[13] Carl Sagan, 'Nuclear war and climatic catastrophe: Some policy implications', *Foreign Affairs* 62:2 (1983), pp. 257–92.

global effects.[14] In a world of much-reduced nuclear weapons—at least a theoretical possibility in the early years of the 21st century—a best-worst-case scenario of 'nuclear autumn' might be possible.[15] But such calculations relied on there being just a handful of nuclear weapons in existence, not the several thousand that exist today.

Effects on the planet would of course be dependent on a number of different variables: the size of the yield/device; whether the detonation was in the air or on the ground (ground bursts cause more fallout/debris), how efficient the fission/fusion process was,[16] and how many warheads were detonated. But in 2022, a study published in *Nature Food* suggested that 'in the most severe scenario' five billion people could die as a result of the likely famine caused by the soot released from a nuclear war between the US and Russia, and possibly two billion from a nuclear war between India and Pakistan.[17] This would be *in addition* to the 50–100 million Indians and Pakistanis killed directly from the weapons themselves.[18] It is worth bearing in mind that India and Pakistan combined have less than 5% of global nuclear stockpiles (and mostly 'smaller' yield bombs').[19]

More recently, nuclear testing and especially the likely effects of nuclear use have become bound up with the idea that humans have created a new period of geological time: the Anthropocene. The word Anthropocene literally means 'the era of humans'—and particularly the era where humans began to have a systemic (mostly negative) impact on the planet. The phrase first reached public consciousness after it was used by the Nobel-prize-winning scientist Paul Crutzen in 2000, to refer to the significance of the

[14] E.g. Seth D. Baum, 'Winter-safe deterrence: The risk of nuclear winter and its challenge to deterrence', *Contemporary Security Policy* 36:1 (2015), pp. 123–48. See also Gregory D. Koblentz, Gigi Kwik Gronvall, Brett Edwards, Martin Furmanski, Sonia Ben Ouagrham-Gormley, & Seth Baum, 'The winter-safe deterrence debate', *Bulletin of the Atomic Scientists*, 18 March–2 April 2015, https://thebulletin.org/roundtable/the-winter-safe-deterrence-debate/.

[15] Adam J. Liska, Tyler R. White, Eric R. Holley, & Robert J. Oglesby, 'Nuclear weapons in a changing climate: Probability, increasing risks, and perception', *Environment: Science and Policy for Sustainable Development* 59:4 (2017), pp. 22–33.

[16] Lower efficiency means more radiative material is left, which (especially if detonated near or on the ground) means more fallout.

[17] Lili Xia, Alan Robock, Kim Scherrer, Cheryl S. Harrison, Benjamin Leon Bodirsky, Isabelle Weindl, Jonas Jagermeyr, Charles G. Bardeen, Own B. Toon, & Ryan Heneghan, 'Global food insecurity and famine from reduced crop, marine fishery and livestock production due to climate disruption from nuclear war soot injection', *Nature Food* 3 (2022), pp. 586–96.

[18] For a discussion of this, see Alan Robock, Owen B. Toon, Charles G. Bardeen, Lili Xia, Hans M. Kristensen, Matthew McKinzie, R.J. Peterson, Cheryl S. Harrison, Nicole S. Lovenduski, & Richard Turco, 'How an India-Pakistan nuclear war could start-and have global consequences', *Bulletin of the Atomic Scientists* 75:6 (2019), pp. 273–9.

[19] Probably about 300 of the estimated global inventories of ~12,000.

period of history where humans had effectively become 'a force of nature'.[20]

While there is some disagreement about when this period started—with some pointing to the start of the industrial revolution in the 18th century and others to the 'great acceleration' in human activity that began in the 1950s. There is clear consensus that the invention of nuclear weapons, and especially high-yield thermonuclear weapons or hydrogen bombs (deriving their power from fusion rather than fission), and the possible impact of nuclear war are central to this new epoch. This was expertly summed up by the United Kingdom (UK) astronomer royal, Lord Martin Rees, in a newspaper interview in 2012,

> Over most of history, threats have come from nature—disease, earthquakes, floods, and so forth. But the worst now come from us. We've entered a geological era called the Anthropocene. This started, perhaps, with the invention of thermonuclear weapons.[21]

As Rens van Munster and Casper Sylvest have noted, these fears can be traced back to the work of Jonathan Schell, and in particular his 1982 book *The Fate of the Earth* where he talked about a 'republic of insects and grass' and 'ecological collapse' that would follow nuclear war.[22] But what perhaps makes today different to previous eras is that the effects of nuclear use have become enmeshed with fears about Anthropogenic climate change, or as some would have it, the 'climate emergency'. Essentially, we now have both a fast (nuclear war) and more gradual (planetary heating as a result of carbon emissions) way to destroy the world.

It is now almost universally accepted that the global climate is changing,[23] the world is getting warmer, and we are likely to experience fluctuations

[20] Erle C. Ellis, *Anthropocene: A very short introduction* (Oxford: Oxford University Press, 2018), pp. 1–2.

[21] Martin Rees, 'Cambridge University's "Terminator studies" department—do we really need it?', *The Guardian*, 26 November 2012, https://www.theguardian.com/education/shortcuts/2012/nov/26/cambridge-university-terminator-studies.

[22] Rens van Munster & Casper Sylvest, 'Nuclear weapons, extinction, and the Anthropocene: Reappraising Jonathan Schell', *Review of International Studies* 47:3 (2021), pp. 294–310. As Schell put it, 'They grew out of history, yet they threaten to end history. They were made by men, yet they threaten to annihilate man. They are a pit into which the whole world can fall-a nemesis of all human intentions, actions and hopes. Only life itself, which they threaten to swallow up, can give the measure of their significance'. Jonathan Schell, *The fate of the earth* (London: Jonathan Cape, 1982), p. 3.

[23] Though it is important to note that climate scepticism remains. These views can often be grouped into scientific uncertainty, questioning the human role, and doubt about the overall impact. See Willem Van Rensburg, 'Climate change scepticism: A conceptual re-evaluation', *SAGE Open* 5:2 (2015).

in weather patterns and possibly more extreme climate events. The leading authority on this is the Intergovernmental Panel on Climate Change (IPCC),[24] which was established by the UN in 1988 to assess scientific research on climate change and make predictions about the future. This was followed in 1992 by the UN Framework Convention on Climate Change (UNFCCC)—an international treaty to combat climate change, which was updated with the 1997 Kyoto Protocol[25](which set emissions targets for *developed* countries) and the 2015 Paris Agreement (committing *all* countries to limiting global warming).[26]

IPCC reports play a key role in shaping the discussions at the UNFCCC. The most recent IPCC report (2023), warned that, 'Climate change is a threat to human well-being and planetary health (*very high confidence*). There is a rapidly closing window of opportunity to secure a liveable and sustainable future for all (*very high confidence*)'.[27] Notably, the report also highlighted that:

> Human activities, principally through emissions of greenhouse gasses, have unequivocally caused global warming, with global surface temperatures reaching 1.1°C above 1850–1900 in 2011–2020.[28]

Essentially, and at the risk of oversimplifying, human activities, especially the burning of fossil fuels, have pumped masses of carbon into the atmosphere, which in turn traps heat and leads to planetary warming and other climatic effects. A different—and slightly mind-boggling—way of conceptualising this is provided by science writer Joseph Romm in *Climate Change: What Everyone Needs to Know*:

> The rate of heat build-up for the entire planet is 250 trillion watts, a large and abstract number. A number of scientists have made the analogy to the energy released by the Hiroshima atomic bomb. By that metric, the current rate of increase in global warming is roughly the same as detonating 400,000 Hiroshima bombs per day, 365 days per year.[29]

[24] https://www.ipcc.ch.
[25] https://unfccc.int/kyoto_protocol.
[26] https://unfccc.int/process-and-meetings/the-paris-agreement.
[27] 'Climate change 2023 synthesis report: A summary for policymakers', Intergovernmental Panel on Climate Change (2023), https://www.ipcc.ch/report/ar6/syr/downloads/report/IPCC_AR6_SYR_SPM.pdf.
[28] 'Climate change 2023 synthesis report', p. 4.
[29] In Joseph Romm, *Climate change: What everyone needs to know* (Oxford: Oxford University Press, 2022), p. 24.

In 2024, the UN General Assembly agreed to establish a high-level scientific panel to study the 'The climatic, environmental, and radiological effects of nuclear war and how these effects impact public health, global socioeconomic systems, agriculture, and ecosystems'.[30] While this will provide a capstone for work that has been ongoing since at least the 1980s, the findings are likely to provide a stark (re)warning about the possible impact of nuclear technologies on our planet and on humans as a species. This in turn is likely to invigorate those pushing for nuclear disarmament and nuclear fairness.

The argument here is not so much that our ability as a species to destroy the planet through nuclear use has changed as we enter the Third Nuclear Age—we (or more specifically the leaders of a handful of nuclear-armed countries) have probably been able to do this since the late 1950s—but rather that the understanding of this has become much more widespread and accepted—and decried by the 95% of states and several billion people that don't have the ability to blow up the earth. It is also because nuclear challenges have become interlinked with debates about global climate change and climate justice, sustainable development, and the wider politics of the Anthropocene. These increasingly act as threat multipliers.

A nuclear panacea?

After the Obninsk reactor in Russia demonstrated the feasibility of providing electricity to the power grid in June 1954, Queen Elizabeth II opened the world's first full-scale nuclear powerplant at Windscale in the English Lake District in October 1956.[31] Despite the fact that the Calder Hall reactor caught fire in 1957 (releasing significant amounts of radioactive material), this marked the beginning of a rapid expansion of nuclear powerplants across the developed world. In the 1960s, nuclear power reactors were built in the UK, US, Canada, France, Sweden, Japan, and Russia, laying the foundations for a 'boom' period in the 1970s. According to the World Nuclear Industry Status Report, at its peak in 1979 there were 236 nuclear reactors

[30] 'Nuclear war effects and scientific research', Resolution 79/238, adopted by the General Assembly on 24 December 2024, https://documents.un.org/doc/undoc/gen/n24/427/04/pdf/n2442704.pdf.

[31] While built principally for electricity production, the plant was also designed to produce plutonium for the UK's nuclear bomb programme. For an overview, see Laura Arnold, *Windscale 1957: Anatomy of an accident* (Basingstoke: Palgrave MacMillan, 2007).

under construction across the world. This reduced markedly over the next few decades (especially after the 1986 Chernobyl disaster in Ukraine), until a mini resurgence began in the 2010s.[32]

As of 2025, nuclear energy provides just under 10% of the world's electricity from about 420 power reactors in 32 countries. This equates to about one-quarter of the world's low-carbon power generation. The vast majority of these facilities remain in the developed world, and concentrated in the US, France, China, Russia, India, Japan, and South Korea.[33] Many states that currently operate nuclear powerplants are in the process of replacing or building new reactors—ensuring that nuclear power will continue to play a key role in energy generation for decades to come.[34] China continues to lead the way—aiming for 150 new reactors by 2035.[35] But the so-called 'nuclear revival' in the US announced by the Trump administration in 2025 is also indicative of this renewed trajectory.[36] As is the announcement in 2023 that the Swedish government will begin a significant reinvestment in nuclear energy that will see nuclear capacity almost double by the 2040s.[37] In addition to nuclear power generation, over 50 countries utilize nuclear technology in about 220 research reactors (used for medical or other scientific purposes).[38]

[32] World Nuclear Industry Status Report, '70+ years of nuclear power: World nuclear power reactors 1951–2025. The WNISR Interactive DataViz, 6 March 2026, https://www.worldnuclearreport.org/reactors.html#tab=iso.

[33] As of 2025, the US leads the world in nuclear reactors, operating 94 in total. France follows with 56 reactors, while China has 54. Russia operates 37 reactors, and Japan has 33. South Korea maintains 25, while India has 22, and Canada runs 19. Ukraine has 15 reactors, while the UK has 9. Spain and Belgium each have 7, while Sweden and the Czech Republic each have 6. Pakistan has 6 reactors, and Finland has 5. Switzerland runs 4, and Hungary and Slovakia each operate 4. Germany operates 3 reactors, while Argentina operates 3, and the United Arab Emirates runs 3. Bulgaria, Brazil, South Africa, Mexico, and Romania each have 2 reactors, and the Netherlands, Iran, Armenia, and Slovenia each have 1.

[34] According to World Nuclear as of 2025, China has 30 reactors under construction, Russia 7, India 6, Japan 2, South Korea 2, Ukraine 2, the UK 2, and Pakistan 1. https://world-nuclear.org/information-library/current-and-future-generation/nuclear-power-in-the-world-today.

[35] Genevieve Donnellon-May, 'Power China's nuclear ambitions', *The Diplomat*, 20 September 2022, https://thediplomat.com/2022/09/powering-chinas-nuclear-ambitions/.

[36] Ariel Cohen, 'Trumps nuclear revival: Playing catch up', Forbes, 26 May 2025, https://www.forbes.com/sites/arielcohen/2025/05/26/trumps-nuclear-revival-playing-catch-up/.

[37] 'Sweden passes law to fund new generation of nuclear reactors', Reuters, 21 May 2025, https://www.reuters.com/business/energy/swedish-parliament-backs-financing-bill-new-nuclear-power-2025-05-21/.

[38] This includes medical isotope production, materials testing, neutron activation analysis, education/training, and basic scientific research. World Nuclear, https://world-nuclear.org/information-library/non-power-nuclear-applications/radioisotopes-research/research-reactors#:~:text=Many%20of%20the%20world's%20nuclear,reactors%20operating%2C%20in%2053%20countries.

The two nations to buck this trend are Germany and Taiwan. In June 2011 the then German Chancellor Angela Merkel announced the decision to phase out nuclear power by 2023, opting instead to concentrate on developing a greater proportion of energy from renewables.[39] Likewise, the Democratic Progressive Party oversaw the closure of Taiwan's last nuclear power reactor in May 2025, ending five decades of nuclear operations.[40] The Japanese government made a similar decision to phase out nuclear power following the Fukushima nuclear accident in 2011, but in 2025 this was overturned as part of its Strategic Energy Plan.[41] Both the Belgian and Dutch governments have also recently 'recommitted' to nuclear power,[42] the government in Denmark is reconsidering its 40-year ban on nuclear power,[43] and the same is true in Serbia (after 35 years).[44]

Many different indicators suggest that nuclear technologies will become more rather than less important in the Third Nuclear Age. This includes extra investment by current nuclear operators as well as new state and non-state aspirants—presenting different types of challenges.[45] Much of this renewed interest can be linked with reducing greenhouse gas emissions and pursuing sustainable development.[46] As International Atomic Energy

[39] See 'The nuclear phase out in Germany', Federal Office of the Safety of Nuclear Waste Management, https://www.base.bund.de/en/nuclear-safety/nuclear-phase-out/nuclear-phase-out_content.html.

[40] See Philip Hou, 'Taiwan's self-created Achilles' heel: Eliminating nuclear power', *The Diplomat*, 16 May 2025, https://thediplomat.com/2025/05/taiwans-self-created-achilles-heel-eliminating-nuclear-power/.

[41] Under this plan, Japan aims to produce 20% of energy from nuclear by 2040. 'The 7th strategic energy plan', Agency of Natural Resources and Energy, February 2025, https://www.enecho.meti.go.jp/en/category/others/basic_plan/pdf/7th_outline.pdf.

[42] Anne-Sylvaine Chassany & Alice Hancock, 'Germany drops opposition to nuclear power in rapprochement with France', *Financial Times*, 19 May 2025, https://www.ft.com/content/e99efa2b-338a-4065-89c6-0683d5759ed7.

[43] Jillian Ambrose, 'Denmark rethinking 40-year nuclear power ban amid Europe-wide shift', *The Guardian*, 14 May 2025, https://www.theguardian.com/world/2025/may/14/denmark-rethinking-40-year-nuclear-power-ban-amid-europe-wide-shift.

[44] 'Serbia plans for future with nuclear energy', *World Nuclear News*, 5 December 2024, https://www.world-nuclear-news.org/articles/serbia-plans-for-future-with-nuclear-energy.

[45] See Richard K. Lester & Robert Rosner, 'The growth of nuclear power: Drivers and constraints', *Daedalus* 138:4 (2009), pp. 19–30.

[46] Though it is possible that some interest in nuclear power in Europe could be for other security reasons, e.g. Maha Siddiqui & M.V. Ramana, 'Eastern Europe's purchase of US nuclear reactors is primarily about military ties, not climate change', *Bulletin of the Atomic Scientists*, 2 August 2024, https://thebulletin.org/2024/08/eastern-europes-purchase-of-us-nuclear-reactors-is-primarily-about-military-ties-not-climate-change/?utm_source=Newsletter+&utm_medium=Email+&utm_campaign=ThursdayNewsletter08082024&utm_content=NuclearRisk_EasternEuropeNuclearReactors_08022024.

Agency (IAEA) Director Raphael Grossi explained in a 2023 address to the IAEA General Conference:

> Seventy years after US President Dwight D. Eisenhower gave his famous speech to the United Nations General Assembly, his Atoms for Peace vision has become an important, active part of our contribution towards international peace and security and furthering sustainable development.[47]

This was echoed by a 2022 report from the International Energy Agency which declared that: 'Nuclear energy plays a significant role in a secure global pathway to net zero'. And that 'Less nuclear power would make net zero ambitions harder and more expensive'.[48] There is even a growing acceptance amongst some environmental scientists that 'safe' nuclear must be part of the solution.[49]

At the time of writing, according to World Nuclear, 30 countries are 'considering, planning or starting nuclear power programmes ... and a further 30 have expressed an interest'.[50] This includes, Bangladesh, Egypt, and Turkey who all have their first reactors under construction; Jordan, Kazakhstan, Lithuania, Poland, Saudi Arabia, and Uzbekistan who have specific programmes for nuclear powerplants in place; and many others including Algeria, Azerbaijan, Estonia, Ethiopia, Ghana, Kenya, Indonesia, Laos, Morocco, Nigeria, Philippines, Rwanda, Sri Lanka, Sudan, Thailand, and Vietnam who have expressed a strong interest in nuclear power.[51] The vast majority of these programmes are in coordination with the Russian state-owned company Rosatom.[52] Indeed, this has led to some concern

[47] 'Statement to the sixty-seventh regular session of the IAEA general conference', 25 September 2023, https://www.iaea.org/newscenter/statements/statement-to-the-sixty-seventh-regular-session-of-the-iaea-general-conference.

[48] 'Nuclear power and secure energy transitions: From today's challenges to tomorrow's clean energy systems', *International Energy Agency*, September 2022, p. 7, 8, https://iea.blob.core.windows.net/assets/016228e1-42bd-4ca7-bad9-a227c4a40b04/NuclearPowerandSecureEnergyTransitions.pdf.

[49] Thom Patterson, 'Time to go nuclear', *CNN World*, 3 November 2013, https://edition.cnn.com/2013/11/03/world/nuclear-energy-climate-change-scientists/index.html.

[50] 'Nuclear energy countries', World Nuclear, https://world-nuclear.org/information-library/country-profiles/others/emerging-nuclear-energy-countries.aspx#:~:text=About%2030%20countries%20are%20considering,their%20first%20nuclear%20power%20plants.

[51] 'Nuclear energy countries'.

[52] E.g. Nick Ferris, 'Russia is pushing to take control of Africa's nuclear future', *The Independent*, 13 May 2025, https://www.independent.co.uk/news/world/africa/russia-nuclear-power-trump-aid-africa-b2741038.html#. See also 'Opening new horizons for cooperation', Rosatom Newsletter #262, February 2023, https://rosatomnewsletter.com/2023/03/02/opening-new-horizons-for-cooperation/

in the Western world that a nuclear energy arms race may be underway, and that organisations such as China National Nuclear Corporation, China General Nuclear Power Group, as well as Rosatom are 'winning' and gaining significant influence as a result.[53] As energy technology specialist Juzel Lloyd puts in: 'China and Russia are already leveraging their nuclear energy investments abroad to deepen their economic and political influence over the countries buying their technology'.[54] This is further complicated by the allegation that both Russia and China 'have weaker supply-side export restrictions than the United States and other Western democracies and have repeatedly failed to enforce trade control regimes'.[55] Perhaps partly in response, in 2025 the World Bank announced that it would lift its decades old ban and allow loans to fund nuclear energy projects.[56]

Of course, nuclear energy remains contentious, and a strong anti-nuclear lobby consisting of environmental groups such as Campaign for Nuclear Disarmament, Greenpeace, and Friends of the Earth have sought to challenge the retention and expansion of civil nuclear power. Activists point to safety fears—especially in the wake of accidents such as that at Chernobyl in 1986 and more recently at Fukushima in 2011, the largely unresolved issue of nuclear waste disposal, and the attendant problem of weapons proliferation, they also challenge the extent to which nuclear power is really green[57] or economically superior to cleaner renewable technologies such as wind, solar,

#:~:text=With%20partners%20and%20customers%20across%20Latin%20America%2C, cooperation%20and%20plans%20for%20the%20coming%20year.

[53] Daniel Shats & Peter W. Singer, 'The balance of power is shifting among nuclear energy titans', *Defense One*, 5 October 2022, https://www.defenseone.com/ideas/2022/10/balance-power-shifting-among-nuclear-energy-titans/378067/.

[54] Juzel Lloyd, 'The other nuclear race: America is falling behind China and Russia on energy innovation', *Foreign Affairs*, 28 April 2025, https://www.foreignaffairs.com/china/other-nuclear-race?s=EDZZZ005ZX&utm_medium=newsletters&utm_source=fatoday&utm_campaign=The%20Return%20of%20Great-Power%20Diplomacy&utm_content=20250428&utm_term=EDZZZ005ZX.

[55] Jupiter Huang, 'It's getting hot in here: Climate change as a catalyst for nuclear proliferation', *Georgetown Security Studies Review*, 10 June 2025, https://georgetownsecuritystudiesreview.org/2025/06/10/its-getting-hot-in-here-climate-change-as-a-catalyst-for-nuclear-proliferation/.

[56] Jamie Smith & Claire Jones, 'World Bank lifts ban on funding nuclear energy in boost to industry', *Financial Times*, 11 June 2025, https://www.ft.com/content/d80b68ec-3da8-42ea-82ee-4cab22b31a69?accessToken=zwAGN77jwsPQkdPYC2jsPahC6tOC7kyrIrMaaQ.MEQCIB8GWNoKWlLh2jWdVr4IRX7noCNnwYbENi7PpHhleVPfAiB4kDjPIFJZutDZmMMV-_h3EYkD9sX0VEB_Zv_Tx7E_5w&sharetype=gift&token=1fdae867-1055-4f5e-a439-6b4eb47dbd4e.

[57] E.g. 'Nuclear power not the answer to climate change', Campaign for Nuclear Disarmament, n.d., https://cnduk.org/resources/nuclear-power-not-the-answer-to-climate-change/.

or hydro.[58] As the Greenpeace website puts it: 'Nuclear energy has no place in a safe, clean, sustainable future. Nuclear energy is both expensive and dangerous, and just because nuclear pollution is invisible doesn't mean it's clean.'[59] Some national governments, notably in Austria have also committed whole-heartedly to the anti-nuclear agenda.[60] Beyond this, 'green theory' also suggests that state-controlled nuclear technology is a way of maintaining existing (capitalist/consumerist/growth-focussed/statist) ways of living and of valuing the environment (as a resource rather than as something valuable for its own sake).[61] And in some contexts, anti-nuclear sentiment comes from vested economic interests, such as the 'coal lobby' or other fossil fuel providers.[62]

At a very basic level the maths of nuclear power seems straightforward—approximately 80 kilograms of uranium can serve the electricity needs of about one million people for one day. By comparison, 80 kilograms of wood can probably fuel a large campfire for a dozen people for one evening.[63] But calculating the economic cost of nuclear power *vis-à-vis* other options can be tricky and is often dependent upon particular contexts. In general, *building* a nuclear reactor remains a very expensive operation, especially when compared with building a coal or gas plant. It is also more difficult to extract uranium ore and fabricate nuclear fuel. But *operating* costs, the ability to control power flows, and of course the emissions from nuclear are much more competitive and potentially cheaper than fossil fuels. As a 2020 International Energy Agency report noted, 'electricity from the long-term operation of nuclear power plants constitutes the least cost option for low-carbon generation.'[64]

[58] For a discussion, see Dianne Plummer, 'Nuclear v renewables: Which energy source wins the zero-carbon race?' Forbes, 11 February 2025, https://www.forbes.com/sites/dianneplummer/2025/02/11/nuclear-vs-renewables-which-energy-source-wins-the-zero-carbon-race/.

[59] 'Nuclear power is dirty, dangerous and expensive. Say no to nukes', https://www.greenpeace.org/usa/climate/issues/nuclear/#:~:text=Say%20no%20to%20new%20nukes,doesn't%20mean%20it's%20clean.

[60] See Austria chapter in Andrew Futter, Ludovica Castelli, Cameron Hunter, Olamide Samuel, & Francesca Silvestri, *The global Third Nuclear Age: Clashing visions for a new era in international politics* (London: Routledge, 2025).

[61] I am grateful to Steve Cooke for explaining this point.

[62] E.g. 'Exploring who opposes nuclear energy and why', Climate Coalition, https://climatecoalition.org/who-opposes-nuclear-energy/.

[63] Tim Gregory, *Going nuclear: How the atom will save the world* (Oxford: The Bodley Head, 2025), p. 44.

[64] International Energy Agency, 'Projected costs of generating electricity 2020', https://www.iea.org/reports/projected-costs-of-generating-electricity-2020.

The debate *vis-à-vis* renewables is also complex. As sustainability expert Dianne Plumber explains:

> While nuclear energy offers high-capacity, low-carbon baseload power, it is often hindered by long construction timelines, cost overruns, waste issues and decommissioning challenges. Conversely, renewables such as solar and wind continue to experience declining costs and rapid deployment, supported by government subsidies and technological advancements. However, integrating intermittent renewables into the grid requires significant investment in storage and infrastructure.[65]

While nuclear is likely to remain more expensive than most renewables, even when considering 'levelised cost', it does offer other advantages. Notably, nuclear power can be scaled and up and down, which means there is no need for storage, the power is always available when it's needed and it is not impacted in the same way as renewables by geography.

There are three broad categories that perhaps explain the renaissance and resurgence of interest in nuclear technologies in the Third Nuclear Age: the quest for sustainable low- or zero-carbon development, technological and scientific demands, and the desire for 'nuclear fairness'.

As the Anthropogenic realisation that humans are having a negative impact on earth's climate has spread and become increasingly accepted, most governments have engaged with the need to find more sustainable, less harmful, and more secure ways of generating or acquiring energy. In fact, a 2025 ruling by the International Court of Justice suggested that failing to do so might make countries liable for legal action.[66] While the goal of 'net zero' emissions might seem a long way away, the political elites in most states now see making moves towards reducing greenhouse gas emissions as a staple of international diplomacy and national policy. This has led some governments to champion the virtues of nuclear energy as an alternative to burning fossil fuels such as coal and gas, and perhaps complementary to a

[65] Dianne Plummer, 'Power play: The economics of nuclear vs. renewables', Forbes, 12 February 2025, https://www.forbes.com/sites/dianneplummer/2025/02/12/power-play-the-economics-of-nuclear-vs-renewables/.

[66] 'Obligations of states in respect of climate change', International Court of Justice, 23 July 2025, https://www.icj-cij.org/sites/default/files/case-related/187/187-20250723-pre-01-00-en.pdf.

range of 'renewables', such as wind, solar, or water.[67] As Director of the IAEA Raphael Grossi argued in a July 2024 article,

> It is increasingly clear that nuclear energy must be a central part of getting the world to net-zero carbon emissions ... remedying the ongoing inequality of access will strengthen the regime that prevents the proliferation of nuclear weapons while widening support ...[68]

Indeed, at the UN Climate Change Conference (COP28) in Dubai in late 2023, delegates from 20 countries from four different continents agreed to triple nuclear energy production by 2050.[69] Reducing the emissions from energy generation also fits with the Sustainable Development Goals adopted unanimously by the UN in 2015.[70]

Interest in nuclear power has also been revived by concerns about energy security. This is particularly acute for states that do not have significant fossil fuel resources (or obvious natural advantages that might support renewable energy generation), and who therefore have to rely upon international markets, and often on just a few suppliers for energy. While this isn't a new phenomenon, as the oil crises of the past show, the vulnerability this creates was highlighted for a number of European states in 2022 as Russia threatened to cut off gas supplies in the wake of the invasion of Ukraine. In response, the UK government made investing in nuclear power a renewed priority, and the European Commission launched the 'REPowerEU Plan'. In a 2025 speech, European Union (EU) Commissioner Dan Jorgensen remarked that:

> It is not a sustainable situation that can be justified over time for our continent to rely on energy imports as heavily as we do. We need a strategy to

[67] As Tim Gregory put it, 'If we want to a get to net zero at all, then we have no choice but to decarbonise our energy supply; if we want to achieve it by 2050, then we must do it quickly; and if we want to achieve it without replacing climate change with another set of problems, then we have to do it smartly. To make a success of net zero in a warming world, cool heads must prevail'. Gregory, *Going nuclear*, pp. 51–2.

[68] Rafael Mariano Grossi, 'Nuclear must be part of the solution: Reinforcing the bargain that strengthens security while expanding peaceful use', *Foreign Affairs*, 18 July 2024, https://www.foreignaffairs.com/united-states/nuclear-must-be-part-solution.

[69] 'At COP28, countries launch declaration to triple nuclear energy capacity by 2050, recognizing the key role of nuclear energy in reaching net zero', 1 December 2023, https://www.energy.gov/articles/cop28-countries-launch-declaration-triple-nuclear-energy-capacity-2050-recognizing-key.

[70] E.g. IAEA, 'Nuclear technology for a sustainable future', June 2012, https://sdgs.un.org/sites/default/files/publications/2267iaeario0612.pdf.

eliminate the imports of all fossil fuels, whatever their origin. It will make us safer and more resilient.[71]

While the inclusion of nuclear power—which contributes over 20% of EU energy requirements—has historically been contentious,[72] the decision taken by the German government in 2025 to drop its opposition to nuclear power in EU energy strategy may well lead to a greater role for nuclear power on the continent going forward.[73]

Nuclear technologies are also experiencing a renaissance due to their perceived value for high-tech, immensely power-hungry industrial, technological, and scientific demands. The idea that nuclear energy might provide a useful power source for space exploration can be traced back to Project Orion in the 1950s as US scientists examined building a nuclear rocket for space exploration and a nuclear-powered spaceship (as well as nuclear-powered missiles).[74] Most nuclear applications in space today (such as the Voyager probes launched by National Aeronautics and Space Administration (NASA) in 1977 to explore the outer regions of our solar system) use radioisotope power systems that generate electricity for small spacecraft through the heat produced by radioactive decay.[75] But the idea of using nuclear fission for propulsion has never gone away; this is partly because nuclear power generation (thermal, electric, or pulse) potentially offers considerable benefits over chemical propellants. The same is true for mini-reactors that might be used for long-term power for space exploration and even settlement. Indeed, in 2023, the UK Space Agency allocated £2.9 million in funding to Rolls Royce to build a nuclear reactor for moon

[71] 'Opening statement of Commissioner Jørgensen in the European Parliament plenary debate on accelerating the phase-out of Russian gas and other Russian energy commodities in the EU', 12 March 2025, https://ec.europa.eu/commission/presscorner/detail/en/speech_25_764.

[72] Kate Abnett & Julia Payne, 'EU countries split over support for nuclear energy', Reuters, 4 March 2024, https://www.reuters.com/sustainability/climate-energy/eu-countries-split-over-support-nuclear-energy-2024-03-04/.

[73] Chassany & Hancock, 'Germany drops opposition to nuclear power'.

[74] For a detailed discussion of the US nuclear space propulsion programme, see Jim Howe, 'Atomic space: A history and analysis of U.S. policy for the development of nuclear space propulsion and power technologies', Unpublished PhD thesis, University of Leicester, November 2023, https://figshare.le.ac.uk/articles/thesis/Atomic_Space_A_history_and_analysis_of_U_S_policy_for_the_development_of_nuclear_space_propulsion_and_power_technologies/24440893?file=42906361.

[75] 'Fact sheet: The Voyager planetary mission', NASA, https://science.nasa.gov/mission/voyager/fact-sheet/.

exploration,[76] and there are plans underway to build a nuclear powerplant on the moon as part of the US-led but multinational Artemis Programme.[77] Concerns about accidents remain, such as the fate of the Kosmos 954 Russian nuclear-powered satellite which crashed to the ground in northern Canada in 1978, but nuclear energy seems likely to be central to what some have termed 'the new space age'.[78] Indeed, nuclear power was essential to Kim Stanley Robinson's compelling vision of life on Mars in his highly acclaimed 1992 futuristic sci-fi novel *Red Mars*.[79]

Back on *terra firma*, nuclear technologies have also been cited as a possible way to provide the enormous, continuous power required by the most sophisticated computer systems, especially those used for data centres, artificial intelligence (AI), large language models, and quantum computing. As Sam Altman, chief executive officer (CEO) of OpenAI put it in 2023, 'I don't see a way for us to get there without nuclear.'[80] In 2024, Microsoft announced that it was planning to reopen the Three Mile Island nuclear powerplant in Pennsylvania (the plant that famously suffered a meltdown in the 1970s) to power a data centre.[81] The same year, Google entered into an agreement with Kairos Power to procure 500 megawatts of electricity from six to seven small modular reactors (SMRs), with the first expected to be operational by 2030.[82] And in October 2024, Amazon announced a $500 million investment in X-energy, a company specializing in advanced modular nuclear reactors.[83] As Matt Garman, CEO of Amazon Web Services,

[76] 'UK Space Agency backs Rolls-Royce nuclear power for moon exploration', UK Space Agency, 17 March 2023, https://www.gov.uk/government/news/uk-space-agency-backs-rolls-royce-nuclear-power-for-moon-exploration.

[77] 'Why NASA wants to put a nuclear power plant on the moon', *CNBC*, 15 November 2200, https://www.cnbc.com/2020/11/15/why-nasa-wants-to-put-a-nuclear-power-plant-on-the-moon.html.

[78] E.g. 'The new space age', *The Washington Post*, 2023, https://www.washingtonpost.com/technology/interactive/2023/new-space-age/.

[79] Kim Stanley Robinson, *Red Mars* (New York: HarperCollins Publishers, 1992).

[80] Catherine Clifford, 'Sam Altman explains why he's helping to take nuclear microreactor company Oklo public via SPAC', *CNBC*, 11 July 2023, https://www.cnbc.com/2023/07/11/sam-altman-talks-about-oklo-nuclear-microreactor-spac.html.

[81] Dawn Stover, 'AI goes nuclear', *Bulletin of the Atomic Scientists*, 19 December 2024, https://thebulletin.org/2024/12/ai-goes-nuclear/?utm_source=ActiveCampaign&utm_medium=email&utm_content=Best%20of%202024%20nuclear%20coverage&utm_campaign=20250106%20Monday%20Newsletter&vgo_ee=nph25wWzCU0CT7nF56CcCG%2FJ3ssdmxjDkH6Cau%2BOmWs8Dn2hRITuSUU%3D%3Av4qUzZKoatSE7OO8X4ChSCINJBuhUYGK.

[82] Michael Terrell, 'New nuclear clean energy agreement with Kairos Power, Google, 14 October 2024, https://blog.google/outreach-initiatives/sustainability/google-kairos-power-nuclear-energy-agreement/.

[83] 'Amazon signs agreements for innovative nuclear energy projects to address growing energy demands', Amazon, 16 October 2024, https://www.aboutamazon.com/news/sustainability/amazon-nuclear-small-modular-reactor-net-carbon-zero.

explained, 'One of the fastest ways to address climate change is by transitioning our society to carbon-free energy sources, and nuclear energy is both carbon-free and able to scale—which is why it's an important area of investment for Amazon.'[84] Interestingly, Chinese tech companies appear less interested (or able) to pursue this route—although Alibaba has apparently invested in Nova Fusion.[85]

Finally, a significant structural driver of interest in nuclear technologies is a feeling of historical redistributive injustice felt by many in the developing world that they have not been able to 'reap the benefits' of nuclear technologies. This is linked with a broader desire for nuclear emancipation and to escape what is sometimes viewed as 'nuclear paternalism' and even neo-colonialism. This goes back to the establishment of an inalienable right to access nuclear technologies enshrined in Article IV of the NPT, but also the founding of the Non-Aligned Movement at the Bandung Conference (in Indonesia) in 1954. Essentially, many prominent voices in 'developing' states believe that the developed world has not fairly shared nuclear technology, and instead has used proliferation and security concerns to prevent access.[86] This has meant that much of the economic development and scientific progress linked with nuclear technology has occurred in the developed rather than developing world. At the same time, as the University of Arizona's Itty Abraham and others have noted, there is also a certain amount of prestige that appears to come with 'mastering' and operating nuclear technologies as a sign of modernity.[87]

What all of this suggests is that the allure of the constant, scalable, long-term supply of power from nuclear fission is here to stay. In theory, fission power, and perhaps in the future fusion power,[88] could play a significant role

[84] Jasmin Jesson, 'Amazon: Nuclear energy for cleaner AI and data centres', *Sustainability Magazine*, 4 November 2024, https://sustainabilitymag.com/articles/amazon-addresses-energy-demands-with-nuclear-energy-projects.

[85] 'Big tech strategic investments show renewed activity in focus?', *Moomoo*, https://www.moomoo.com/news/post/56736697/big-tech-strategic-investments-show-renewed-activity-in-focus?level=1&data_ticket=1759742267775958.

[86] William Potter & Gaukhar Mukhatzhanova, *Nuclear politics and the Non-Aligned Movement* (London: Routledge for the International Institute for Strategic Studies, 2012), pp. 83–4.

[87] E.g. Itty Abraham, *The making of the Indian atomic bomb: Science, secrecy and the postcolonial state* (Zed Books, 1998); and Itty Abraham (ed.), *South Asian cultures of the bomb: Atomic publics and the state in India and Pakistan* (Bloomington, IN: Indiana University Press, 2009).

[88] As Charles D. Ferguson notes, 'An oft-told joke in energy research is that nuclear fusion is the energy source of the future and always will be.' Charles D. Ferguson, *Nuclear energy: What everyone needs to know* (Oxford: Oxford University Press, 2011), p. 17.

in driving the planet away from greenhouse gasses and providing the energy that a burgeoning global population will need to develop sustainably. At the same time, a move towards SMRs will potentially provide greater flexibility, enhanced safety, and lower upfront capital investment.[89]

Other possible uses of nuclear technologies, such as the idea of seeking to 'nuke a hurricane' purportedly raised by President Donald Trump in 2019,[90] destroying asteroids on a collision course with earth (a Hollywood staple),[91] for 'plowshare projects' such as large-scale earth moving,[92] and other forms of 'nuclear geoengineering' as a means to combat climate change are, fortunately, at the absolute margins of the debate.[93] At least for now ...

Towards a world of interconnected risks

The increased commingling of nuclear and climate risks and the possibility of civilian nuclear proliferation is drawing our attention back to a number of challenges associated with nuclear security and safety.[94] While nuclear-climate risks have been around for some time, the changing climate is now also creating a set of entirely new challenges for nuclear operations: and a co-constitutive relationship with the climate is something that is increasingly a Third Nuclear Age phenomenon. Essentially, more nuclear facilities being operated by more actors—both state and non-state—in different geographic and political environments, amplifies the challenges of

[89] Or even 'micro-reactors'. See 'What are small modular reactors (SMRs)?' IAEA, https://www.iaea.org/newscenter/news/what-are-small-modular-reactors-smrs.

[90] Jonathan Swan & Margaret Talev, 'Scoop: Trump suggested nuking hurricanes to stop them from hitting US', *AXIOS*, 25 August 2019, https://www.axios.com/2019/08/25/trump-nuclear-bombs-hurricanes.

[91] David A. Koplow, 'Planetary defense: The nuclear option against Asteroids', *Arms Control Today*, April 2004, https://www.armscontrol.org/act/2024-04/features/planetary-defense-nuclear-option-against-asteroids.

[92] Such as Project Plowshares. See Scott Kaufman, *Project plowshare: The peaceful use of nuclear explosives in cold war America* (Ithaca, NY: Cornell University Press, 2017); and Mllo D. Nordyke, 'The Soviet program for peaceful uses of nuclear explosions', *Science & Global Security* 7:1 (1998), pp. 1–117.

[93] For an interesting discussion of geoengineering, see: James Fleming, *Project Argus. Fixing the sky: The checkered history of weather and climate control* (New York: Columbia University Press, 2012).

[94] With reference to this section's heading, research suggests that climate and nuclear risks have historically been compartmentalised. See Sterre van Buuren, Thomas Fraise, & Benoit Pelopidas, 'Existential silos: The compartmentalization of the futures of environmental change and the nuclear threat', *Futures*, 173 (October 2025).

nuclear safety, security, and non-proliferation. At the same time, the changing climate, including rising sea levels and more extreme weather events will directly impact nuclear facilities and installations, and potentially create new flash points between nuclear-armed actors.

The first concern that is almost always associated with the proliferation of civil nuclear technologies to more actors is their inherent dual use nature, which raises the risk of these technologies being used in nuclear weapons programme.[95] For sure, the historical record is marked by instances where—as Texas A&M's Matthew Fuhrman puts it—atoms for peace can unintentionally become atoms for war,[96] but the path to weaponisation is more nuanced than is often understood. Powerplants or research facilities themselves do not necessarily present a direct weapons proliferation threat, but rather it is the ability to produce or acquire nuclear fuel, and how this is safeguarded, that is important. Most states with nuclear facilities have signed up to the IAEA Additional Protocol that significantly increases the ability of international authorities to monitor nuclear programmes, and most states don't have the ability to enrich uranium or breed or reprocesses plutonium (they buy it from international markets). This system isn't perfect, but should give prior warning if civil nuclear programmes are being diverted.[97] So, the proliferation of civilian nuclear facilities doesn't automatically mean proliferation of nuclear weapons latency—whereby states can move quite close to acquiring the bomb without breaking international law.[98] The proliferation risk comes from cases where the state also develops an indigenous capability to control the nuclear fuel cycle (and the ability to hide this from prying eyes). What makes this complicated is that the right to produce nuclear fuel is protected by the NPT. This legal peculiarity is at the heart of the challenge posed by the Iranian nuclear programme.

But weapons proliferation is not the only risk posed by the diffusion of nuclear technologies. In 1984, US academic Bennett Ramberg published a

[95] See Steven E. Miller & Scott Sagan, 'Nuclear power without nuclear proliferation?', *Daedalus*, 138:4 (Fall 2009), pp. 7–18.

[96] Matthew Fuhrman, *Atomic assistance: How 'atoms for peace' programmes cause nuclear insecurity* (Ithaca, NY: Cornell University Press, 2012), p. 3.

[97] For example, the IAEA released a report in early June 2025 detailing Iran's non-compliance with its nuclear safeguard's operations. Shortly afterwards Israel and the US launched a military attack. Francois Murphy, 'IAEA board declares Iran in breach of non-proliferation obligations', Reuters, 12 June 2025, https://www.reuters.com/world/china/iaea-board-declares-iran-breach-non-proliferation-duties-diplomats-say-2025-06-12/.

[98] The best example being Japan, and increasingly Iran. E.g. Mark Fitzpatrick, *Asia's latent nuclear powers: Japan, South Korea and Taiwan* (London: Routledge for the International Institute for Strategic Studies, 2016).

book called *Nuclear Powerplants as Weapons for the Enemy*, in which he clearly outlined how nuclear facilities could and likely would be targeted by adversaries in times of conflict. While the consequences of a nuclear war had by this time been fairly well documented, far less attention had been given to non-weapons-related nuclear-climatic effects. Nevertheless, the result of an attack on a powerplant, as he explained, could be devastating for both humans and the environment.[99] The catastrophic meltdown of the Chernobyl nuclear powerplant in northeastern Ukraine just two years later—albeit by accident rather than as a result of an attack—and the mass release of radiation this caused began to give some idea of what this might look like.[100]

Nearly four decades later, Ramburg's warning came into stark reality as fears grew that the nuclear facilities at the Zaporizhia nuclear powerplant, this time in southeastern Ukraine, might be targeted as part of the ongoing war with Russia. As former colleagues of mine, Ludovica Castelli and Olamide Samuel have documented, this was not the first example of fears of nuclear facilities being targeted, and indeed, the bombing of the Osirak reactor in Iraq in 1981 and the Stuxnet cyber attack on the Natanz enrichment facility in Iran in 2010 did exactly that. But the possibility of more nuclear facilities being built in or near war zones, in unstable regions, or being considered as strategic targets is becoming more acute. Moreover, as Castelli and Samuel explain, there is currently no international agreement that prohibits such attacks (when such facilities are suspected of having a military dimension).[101] Nuclear installations also potentially make attractive targets for terrorists.[102] While most modern nuclear reactors are well protected against many physical threats—allegedly some reactors can even withstand a direct hit from an aircraft—many auxiliary facilities could be compromised or attackers could find other ways to cause a meltdown and radiological release.[103] Moreover, as the 1979 film *The China Syndrome*—released just

[99] Bennett Ramberg, *Nuclear powerplants as weapons for the enemy* (Oakland, CA: University of California Press, 1984).

[100] Perhaps the best way to engage with this is by watching the mini-series 'Chernobyl' produced by HBO and Sky. For those that prefer an engaging read, see Serhii Plokhy, *Chernobyl: History of a tragedy* (London: Penguin, 2019).

[101] Ludovica Castelli & Olamide Samuel, 'Justifying attacks on nuclear facilities', *The Nonproliferation Review* 30:1–3 (2023), pp. 83–105.

[102] Daniel B. Poneman, *Double jeopardy: Combating nuclear terror and climate change* (Cambridge, MA: MIT Press, 2019).

[103] Avital Wagner, 'Countering threats to nuclear powerplants', *RUSI*, 19 November 2007, https://www.rusi.org/publication/countering-threats-nuclear-power-plants#:~:text=In%20the%20US%2C%20nuclear%20power,television%20systems%20and%20alarm%2Falert.

days before the Three Mile Island accident in Pennsylvania—suggested, corporate greed, sloppy protocols, and cover-ups could be equally problematic.[104] It is however important to note the difference between attacking a working nuclear powerplant and attacking a nuclear facility (such as an enrichment cascade or storage bunker). Attacking an operational nuclear powerplant would release potentially massive amounts of radioactive material, whereas attacking other nuclear facilities would be far less acute because the material is stable. This is arguably why the US and Israel attacked Natanz, Fordo, and Isfahan, but did not attack the Bushehr reactor in Iran as part of Operation Midnight Hammer in 2025.

But it is not just the possible future *nuclear impacts on the climate* as nuclear technologies spread that should concern us. A growing body of work has also pointed to the possible *effects of climate change on nuclear installations*, both civilian and nuclear. As Andrea Rezzonico, assistant director of the Climate-Nuclear Security project at the Council on Strategic Risks, points out, several countries are 'exploring nuclear energy development while navigating serious security and climate related threats' and 'Many reactors continue to be constructed along coastlines'.[105] Plans underway in Bangladesh, Nigeria, Egypt, and Turkey are all cited as possible concerns. While the earthquake and tsunami that led to a radioactive release at Fukushima in Japan in 2011 might be considered a freak event, rising sea levels and more extreme whether events raise questions about the safety of many powerplants around the world. Indeed, in 2024 the US Government Accountability Office recommended that the Nuclear Regulatory Commission assess the risk posed by climate projections—suggesting that perhaps this hadn't been a top priority in the past ...[106]

Neither is it just civilian nuclear facilities that might be vulnerable, but the wider nuclear *weapons* ecosystem too. As part of a wide-reaching review of the nuclear-climate literature, Norwegian political scientist Kjølv Egeland made an important point, and something often overlooked in nuclear studies:

[104] See Aljean Harmetz, 'Fallout from "China syndrome" has already begun', *The New York Times*, 11 March 1979, https://www.nytimes.com/1979/03/11/archives/fallout-from-china-syndrome-has-already-begun-the-china-syndrome.html.

[105] Andrea Rezzonico, 'Climate change and nuclear developments: A growing security nexus', *SAIS Review* 39:2 (2019), pp. 141, 145.

[106] 'Nuclear power plants: NRC should take actions to fully consider the potential effects of climate change', GAO-24-106326, Government Accountability Office, April 2024, https://www.gao.gov/assets/gao-24-106326.pdf.

> The existing literature on nuclear strategy is in large measure written on the increasingly untenable assumption that the world beyond the game of nuclear bargaining is a constant—a void or vacuum within which nuclear politics plays out. Climate change has the potential to radically upset this premise.[107]

As the Carnegie Endowment's Jamie Kwong and others have detailed, changes in weather patterns, extreme weather events, and rising sea levels could impact nuclear weapons operations directly: the facilities that house nuclear weapons, such as air bases and naval ports could all potentially be vulnerable. And while it might be possible to move these facilities for some, it would be very costly.[108] As Kwong notes, 'Climate change could have mission-altering impacts on the U.S. nuclear deterrent' due to rising sea levels, flooding, and extreme weather.[109] In particular, the Kings Bay submarine base (in Georgia) could be vulnerable to rising sea levels and flooding; warming temperatures could cause flooding impacting intercontinental ballistic missile (ICBM) silos at Minot air force base (North Dakota), and extreme precipitation could impact the operation of strategic bombers at Whitman air force base (Missouri).[110] Likewise, Allie Maloney warned in a piece for the *Bulletin of the Atomic Scientists* in April 2025 that 'climate change could "make the US submarine force inoperable over the coming decades"'.[111] The wider nuclear weapons ecosystem might also be vulnerable as the wild fires that caused temporary closure of the Pantex nuclear plant in Texas in 2024 demonstrated.[112]

[107] Kjølv Egeland, 'Disentangling the nexus of nuclear weapons and climate change—A research agenda, *International Studies Review* 27:1 (March 2025).

[108] While not for the same reason, a study conducted by Malcolm Chalmers and William Walker into what Scottish independence might mean for the UK nuclear arsenal gives some indication into the difficulties of relocating nuclear facilities. Malcom G. Chalmers & William Walker, *Uncharted waters: The UK, nuclear weapons and the Scottish question* (East Linton: Tuckwell Press, 2001).

[109] Jamie Kwong, 'How climate change challenges the US nuclear deterrent', Carnegie Endowment for International Peace Working Paper, 10 July 2023, https://carnegieendowment.org/research/2023/07/how-climate-change-challenges-the-us-nuclear-deterrent?lang=en.

[110] Kwong, 'How climate change challenges the US', pp. 3–4.

[111] Allie Maloney, 'How climate change could disrupt the construction and operations of US nuclear submarines', *Bulletin of the Atomic Scientists*, 14 April 2025, https://thebulletin.org/2025/04/how-climate-change-could-disrupt-the-construction-and-operations-of-us-nuclear-submarines/?utm_source=ActiveCampaign&utm_medium=email&utm_content=Climate%20change%20and%20nuclear%20submarines&utm_campaign=20250414%20Monday%20Newsletter&vgo_ee=WB1kDVAzSAvn8bRTjj08QNFDFoepExQf4anBbvgpMZc5IRtwxV1bswQ%3D%3A2wLznyYXyfUpOrTmuvdLWEfbAguVAhwd.

[112] Jessica McKenzie & Francois Diaz-Maurin, 'Texas wildfires force major nuclear facility to briefly pause operations', *Bulletin of the Atomic Scientists*, 28 February 2024, https://

Across the Atlantic, the challenge could be even more acute, as Sussex University's Paul Dorfman wrote in the Executive Summary of a report into 'Climate change and UK nuclear': 'UK nuclear is quite literally on the front-line of climate change—and not in a good way'.[113] In addition to the challenges this poses for powerplants along the UK coast, it is conceivable that military bases such as those on the Gare Loch and Loch Long ~45 miles west of Glasgow, where UK nuclear-armed submarines are based and nuclear warheads are housed, could flood too. In August 2025, a wildfire also threatened RAF Fylingdales (home to a US early warning radar) in North Yorkshire.[114] While much less is known about climate risks to nuclear facilities in other countries, in 2010 and 2021 wildfires threatened the Sarov nuclear production facility located in the Nizhny Novgorod Oblast in western Russia.[115] One must assume that climate will impact nuclear operations in the other six nuclear-armed states, and maybe the other countries where nuclear arms are based too.

In 1997, the video game *Fallout* was released to much general excitement among PC gamers (at least according to my step-son Zane, I haven't played it ...). The premise of the game—which was later turned into a TV series for Amazon Prime—was based on a nuclear war caused by future resource scarcity. This link between rapid population growth, dwindling resources—possibly as a result of a changing climate—and an increased likelihood of conflict (possibly between nuclear powers) is the final worrying possible incarnation of the nuclear-climate-development nexus in the Third Nuclear Age and beyond. In this way, *Fallout* and other similar computer games may be a harbinger of how both climate change and developmental demands become a driver of insecurity and conflict between nuclear-armed states as the Third Nuclear Age unfolds.

Unfortunately, the possibility explored in *Fallout* is not as far-fetched as some might hope. As far back as 1798 the economist Thomas Malthus proposed that population growth would one day outstrip the planet's ability

thebulletin.org/2024/02/texas-wildfires-force-major-nuclear-weapons-facility-to-briefly-pause-operations/.

[113] See also Paul Dorfman, 'Climate impact on UK nuclear military', Nuclear Consulting Group, 2021, https://infrastructure.planninginspectorate.gov.uk/wp-content/ipc/uploads/projects/EN010012/EN010012-004907-Stop%20Sizewell%20C%20-%20Summaries%20of%20all%20WRs%20exceeding%201500%20words%207.pdf.

[114] Julia Bryson & Emily Johnson, 'Moorland fire to burn "for a number of days"', *BBC*, 14 August 2025, https://www.bbc.co.uk/news/articles/ce3jkney415o.

[115] 'Wildfires near Russia's nuclear research center spark state of emergency', *The Moscow Times*, 24 August 2021, https://www.themoscowtimes.com/2021/08/24/wildfires-near-russias-nuclear-research-center-spark-state-of-emergency-a74878.

to feed and support us.[116] With a global population estimated at over eight billion in 2025, which could reach 10 billion by 2050, and even if stabilising by 2100, the challenge of providing food, water, and energy will grow exponentially.[117] While technological change and other developments will be important, it is hard to see a future that is not characterised by the need for constant and sustainable access to energy (such as that provided by nuclear), but also place more pressure on opening up new ways to produce food and secure access to water. This in turn may drive conflict, and possibly proliferation.

This also means that climate-induced events could create geostrategic choke points or drive mass migration due to flooding, or lack of food and access to water. The most widely cited example of this risk is how access to water from the Himalayan glaciers presents a possible flashpoint between India and Pakistan.[118] As an upper riparian state, recent threats by India to cut off Pakistani access to water from the Indus River in the aftermath of the 2025 Pahalgam crisis are perhaps indicative of this.[119] As Princeton University's Zia Mian explained in a 2016 article:

> the inevitable pressures from climate change over the coming decades on the Himalayan glaciers, the monsoons, and ground water in the Indus Basin, which will lead to reduced and less reliable access to water in an already water-stressed region, at a time of rapidly growing demand. These drivers have already started to overlap, and conflicts over land, people, blood, and water may become one.[120]

What this suggests is that the Third Nuclear Age will be characterised by interconnected existential risks, with nuclear technologies very much at its heart. This is now an era where the ability to split the atom and

[116] Thomas Malthus, *An essay on the principle of population, as it affects the future improvement of society. With remarks on the speculations of Mr. Godwin, M. Condorcet and other writers* (J. Johnson, 1798).

[117] See van Dijk, M., Morley, T., Rau, M.L. et al., 'A meta-analysis of projected global food demand and population at risk of hunger for the period 2010–2050', *Nature Food* 2 (2021), pp. 494–501.

[118] Sarang Shidore, 'Climate change and the India-Pakistan rivalry', Council on Strategic Risks Briefer No.4, 23 January 2020), https://councilonstrategicrisks.org/wp-content/uploads/2020/02/ShidoreJan23ClimatePakistan.pdf.

[119] Simmone Shah, 'Climate change is straining Pakistan's water. Tensions with India could make it worse', *Time*, 9 May 2025, https://time.com/7284470/india-pakistan-water-supply-climate-change/.

[120] Zia Mian, 'Kashmir, climate change and nuclear war', *Bulletin of the Atomic Scientists*, 7 December 2016, https://thebulletin.org/2016/12/kashmir-climate-change-and-nuclear-war/.

utilise the enormous energy this produces are co-constituted with the expansion of AI, growing socio-economic inequality and resource scarcity, biotechnology, and of course humanmade changes to our planetary climate system.

Conclusion: The climate-nuclear Janus

In his highly influential 1986 book *The Risk Society* (translated into English in 1992), German sociologist Ulrich Beck argued that modern societies are defined by the distribution and management of humanmade global risks, and that the technologies that produced modernisation have now become the source of the greatest challenges that we face.[121] Four decades later, this seems to capture the core dynamics at the heart of the nuclear-climate-development nexus and the broader theory of the Anthropocene. Essentially, the Third Nuclear Age will be shaped by the need to deal with problems that are fundamentally of our own making, and while the effects may be global, they are also likely to exacerbate global inequalities and injustice.

The safe management of nuclear technologies, the health of our planet, and sustainable and fair socio-economic development have been central to the nuclear story since 1945. But in the Third Nuclear Age it seems that these major challenges have become further entwined, indivisible, and co-constitutive. If, as climate science suggests, we are moving closer to an inflection point where our widespread use of fossil fuels as a species has to change, and where exponential demand for energy to meet economic and social developmental needs continues, then nuclear is likely to play some part in this. Not perhaps the most dominant, but for some it will be a major component of moves towards net zero, driving human development and addressing climate justice, and perhaps pushing other scientific and technological frontiers as well.

The Third Nuclear Age may well see a return to explosive nuclear testing—the rejection of which, notwithstanding North Korea—has remained a global norm for nearly 30 years—or well over a third of the

[121] Ulrich Beck, *The risk society: Towards a new modernity* (London: Sage, 1992).

nuclear age.[122] And while impossible to quantify,[123] there is a feeling that the prospect of the deliberate or accidental use of a nuclear weapon is increasing rather than decreasing as states double down on nuclear deterrence, new weapons systems are deployed, and new flashpoints emerge. Ultimately, we still live in a world characterised by the possibility of nuclear winter or anthropogenic destruction as a result of even a 'limited' nuclear war. At the same time, the growing demand for nuclear power to drive development and the spread of nuclear installations to more actors in different parts of the world to meet burgeoning energy and scientific demands can only increase the risk of radiation release because of accidents or the deliberate targeting of these facilities. But this is not a one-way relationship: both military and civilian nuclear facilities could also be compromised by rising sea levels, flooding, and more extreme weather in general.

Managing the climate-nuclear-development nexus won't be easy. It will involve a patchwork quilt of mechanisms designed to enhance reactor safety and security, a system of reliable nuclear fuel supplies, as well as tough decisions about the basing of nuclear weapons and security installations. It may be possible that future fusion reactors or SMRs can be made more proliferation resistant, but it is also difficult to escape the 'original sin'[124] of nuclear technology—that the spread of nuclear capabilities for ostensibly peaceful purposes has often been used to develop weapons programmes, and that this might happen in the future.[125]

Ultimately, radiological release, nuclear use, and nuclear war are distinct possibilities in the Third Nuclear Age, and perhaps more acute today than at any point in our nuclear past given the wider Anthropogenic context within which these challenges will play out.

[122] In 2025, President Donald Trump seemed to suggest that the US might resume nuclear testing, but the actual meaning of this is debated. See, for example, James Acton, 'Trump has an out on nuclear testing. He should take it', *Emissary*, Carnegie Endowment for International Peace, 6 November 2025, https://carnegieendowment.org/emissary/2025/11/trump-nuclear-testing-offramp?lang=en.

[123] For a discussion of this, see Amy J. Nelson & Alexander H. Montgomery, 'How not to estimate the likelihood of nuclear war', Brookings, 10 October 2022, https://www.brookings.edu/articles/how-not-to-estimate-the-likelihood-of-nuclear-war/.

[124] Also, the title of Bleddyn Bowen's excellent book about the history of space. In this case, the 'original sin' of space technology is that it always had a nuclear dimension. Bleddyn Bowen, *Original sin: Power, technology and war in outer space* (London: Hurst Publishers, 2022).

[125] E.g. Fuhrman, *Atomic assistance*; and Matthew Kroenig, *Exporting the bomb* (Ithaca, NY: Cornell University Press, 2011).

Conclusion

'A strange game. The only winning move is not to play'

Short of some revolutionary development in the science of violence that supplants the massive 'bang for your buck' of a nuclear blast, or a cheap, quick, and scalable change in the way that modern societies create 'clean' energy, nuclear capabilities will continue to fulfil important functions for a number of states across the globe.[1] This may not be the news that everyone wants to hear as we enter the Third Nuclear Age, but from the current vantage point it is the most honest and realistic. The result is that whether one likes it or not, nuclear technologies will probably be with us for the foreseeable future, and a nuclear-weapons-free world probably remains a distant dream. Essentially, and while the central cast and headline grabbing themes may have shifted since 1945, the existential realities of living with the bomb remains largely the same.

But the era that we are entering has some notable differences from the past and this necessitates an analysis not just of what has changed and why, but a rethink of our approach to understanding this context. In this way, the Third Nuclear Age is not just about a change in the main focus and locus of global nuclear risks and what worries smartly dressed people at breakfast meetings in the capitals of the most powerful states, it is also intrinsically bound up with the need to take a broader conceptual view of global nuclear politics. The Third Nuclear Age will therefore be both materially and intellectually different from the past. Or as United Nations UN) Secretary General Antonio Guterres warned in a 2020 speech: 'Geopolitical tensions, the climate crisis, global mistrust and the downsides of technology can jeopardise every aspect of our shared future'.[2]

[1] Source for quote in chapter subtitle: Joshua—*WarGames*.

[2] Antonio Guterres, 'Remarks to the General Assembly on the Secretary-General's priorities for 2020', United Nations, 22 January 2020, https://www.un.org/sg/en/content/sg/speeches/2020-01-22/remarks-general-assembly-priorities-for-2020.

The Third Nuclear Age. Andrew Futter, Oxford University Press. © Andrew Futter (2026).
DOI: 10.1093/9780198944126.003.0008

Six key dynamics sit at the heart of the Third Nuclear Age thesis: first, while the way that we conceptualise our nuclear world has always reflected the views of the most dominant actors within the international system, and particularly the views and interests of a handful of political, military, and intellectual elites, nuclear knowledge is slowly becoming more democratised and contested; second, that significant advances in technology are impacting the ways, ends, and means of both nuclear and conventional missions, blurring previous distinctions between weapons systems, and recasting thoughts about strategic stability and escalation, but this is largely evolutionary rather than revolutionary; third, this technological change is affecting both the demand and supply side of nuclear deterrence and proliferation, expanding the methods and approaches being taken to secure against nuclear threats, and in some cases reopening the temptation of nuclear coercion; fourth, while 'great power nuclear competition' is central to the emerging era, this must be understood alongside the interests, fears, and geopolitical dynamics that motivate and concern the other 'bit-part-player' states often seen as more peripheral in the nuclear world; while some middle powers may see a need to rethink nuclear options as security complexes shift, the majority of states will remain non-nuclear-armed and (and least for the time being) wedded to disarmament and alternative pathways to security; fifth, our nuclear world is increasingly characterised by disagreement, division, and in some cases demonisation—this ranges from arms control and the viability and fairness of international nuclear regimes up to very different visions of our imagined nuclear future; sixth and finally, nuclear technologies and their associated risks have become increasingly enmeshed and interconnected with other major challenges to society, but we as a species seem ill prepared to meet the planetary-scale implications of what is increasingly referred to as the Anthropocene.

We have survived life with nuclear technologies for over three-quarters of a century, and we have collectively come up with a range of techno-political, supervisory, regulatory, and normative measures to manage and administer our nuclear world. But this does not mean that we have solved the nuclear problem, or that we have somehow devised enduring solutions. Nuclear deterrence probably has played a role in limiting or preventing conflicts over the past eight decades, but this is far from fool-proof. At the same time, key mechanisms such as the Non-Proliferation Treaty (NPT) and the International Atomic Energy Agency (IAEA), and norms such as the taboo have also played a role in limiting proliferation and nuclear use too—but ultimately these are not guarantees of future restraint. Despite general—sometimes begrudging—acceptance, neither does this mean that the nuclear order that we—or perhaps more correctly, a handful of prominent leaders, officials, and intellectuals—have created is necessarily fair or equitable; it largely reflects the interests of the most powerful actors, particularly the United States (US) and the other nuclear-armed states.

Of course, the reality is that one can create different narratives of the nuclear past to serve particular purposes: whether it be in support of nuclear deterrence and non-proliferation and how this has prevented major war; a warning that at some stage our luck will run out and nuclear use of some type will be inevitable; or that it is still possible to find some balance that suits all of us where the 'saving power' of the atom can be realised. This of course also impacts nuclear learning as it is entirely dependent upon what one chooses to focus on and how that fits with already entrenched world-views. But for most of the nuclear age so far this has been the product of a particular view and experience of global politics and history—one wedded in the post-war US and to a lesser extent the developed, democratic, Western world.

Whether one agrees with the framing of today's nuclear world as a Third Nuclear Age, accepts that the nuclear order is in a considerable state of flux, or sympathises with the need to move beyond the ethnocentric origin and focus of the way that nuclear order is conceived, the key to our nuclear future is an open, honest, good-faith interaction. It is possible to sit with people you disagree with and talk. Even if you can't find solutions, it is highly likely that you will learn something or build a relationship that just might become important in the future. This also means ensuring that we have the correct fora for this type of interaction to happen and of course a diverse range of stakeholders who can participate.[3] We live in an unprecedented world of immediate interconnectedness and masses of instantly accessible information, but in the nuclear realm we seem to be moving further away from each other and from seeking to fully engage with the nuclear condition. This might also mean that we need to revisit how we create, reify, and challenge what constitutes nuclear knowledge more broadly.[4]

The labelling of a Third Nuclear Age is just one way to capture the current period in our 80-year courtship with nuclear technologies. Naturally,

[3] See, for example, Olamide Samuel, 'Travelling while black: A first-hand account of the restrictive visa system impacting diversity at nuclear policy conferences', *European Leadership Network*, 8 August 2022, https://www.europeanleadershipnetwork.org/commentary/travelling-while-black-a-first-hand-account-of-the-restrictive-visa-system-impacting-diversity-at-nuclear-policy-conferences/.

[4] I am grateful to Shivani Singh for this point.

nuclear history looks different depending on where you are in the world, who you are, and what you consider as politically important, but what I hope this book has done is draw our attention to a range of significant changes and dynamics that will reshape how we *all* think about the nuclear age. I have deliberately tried to stand back from what often appears to be an increasingly messy, complex, and multifaceted nuclear world that we inhabit (albeit from my desk in the heart of the nuclear-armed Western world). In doing so, I hope to address my major concern that we are failing to look at the big picture of the nuclear order, often obsessing about the wrong things, and my attendant concern that we may be ill prepared either intellectually or societally to deal with the possibility of a major nuclear shock to the way that we live our lives. In this way, I was quite taken by a quote from Hedley Bull's *The Control of the Arms Race*, published more than 60 years ago, but perhaps equally apposite today:

> This is an inquiry into the modern arms race and the measures by which it might be controlled. It is not a plea for disarmament, or against it, or for any particular military policy. If it contains a plea at all, this is for the recognition of complexity in the moral, military and political issues raised by modern war: for confronting this complexity, rather than turning away from it: for rigorous study and anxious questioning, in place of the pursuit of panaceas.[5]

By no means do I claim to have all the answers to the challenges at the heart of the Third Nuclear Age that seem likely to lay ahead, neither do I have a crystal ball to look into the future and suggest what a possible Fourth or Fifth nuclear age may look like.

But writing this book, and leading the multi-year project that underpinned it have led me to four key reflections that I think will be fundamental to our nuclear future. I don't expect everyone who reads this book to agree with all of these points, but hopefully they will at least spark debate and discussion.

The centrality of power to nuclear politics

Our nuclear world has always reflected—and continues to reflect—the often-unequal and perhaps unfair distribution of material power and

[5] Headley Bull, *Control of the arms race*, (New York, Praeger: 1965).

ideational influence across the global system. This has meant that political, intellectual, and military elites in the US have had a disproportionately large say in what we think of as global nuclear politics, the types of issues, agendas or challenges that are prioritised, and the way that we think about and manage nuclear weapons more broadly. Indeed, the historical framings of a First, Second, and now Third Nuclear Age only really make sense when viewed from shiny office blocks and grand neo-classical buildings in Washington, DC. The point here is not to pass judgement on the relative merits or otherwise of this phenomenon,[6] but rather to acknowledge that it is a reality that has shaped how all of us on the planet have engaged with nuclear technologies since 1945. This dominance might be becoming less prominent as we push further into the 21st century, but as my colleague Shivani Singh put it, 'While there are trends towards nuclear multipolarity, there is still no reason to believe, at least for now, that the locus of power and influence is going to shift away from the US as the Third Nuclear Age unfolds'.

This means that the global nuclear order will remain largely structured around American interests for the foreseeable future. There are several important implications of this: First, this means that the US has a continued vested interest in limiting the proliferation of nuclear weapons to more states—this goes back to European allies in the 1950s and the continued centrality of the NPT today. Second, the US will continue to be a key player in enforcing counterproliferation, that is using military force to prevent nuclear acquisition—the most recent example being the attacks on Iran in 2025 and 2026. Third, global power considerations continue to reflect the fact that non-proliferation has always trumped disarmament and even peaceful uses in US grand strategy. Fourth, the US has historically engaged in 'selective' sharing of its nuclear technology, notably with the United Kingdom (UK) since 1958, with members of North Atlantic Treaty Organization (NATO) since the 1960s, the 'nuclear deal' with India in 2006, and most recently through the Australia-UK-US treaty (AUKUS)—this has often been seen as reflecting a 'double standard', whereby others do not have the same access. Fifth, and arguably since 1949, the US has been trying to escape the straitjacket imposed by the vulnerability to nuclear-armed adversaries in order to deploy power without limits—or at least to seek a perception of invulnerability that existed before 1945. Somewhat ironically, the unwitting

[6] For discussion of this, see Nick Ritchie, 'A hegemonic nuclear order: Understanding the Ban Treaty and the power politics of nuclear weapons', *Contemporary Security Policy* 40.4 (2019), pp. 409–34; and Jan Ruzicka, 'Behind the veil of good intentions: Power analysis of the non-proliferation regime', *International Politics* 55 (2018), pp. 369–85.

or otherwise centrality of nuclear weapons for the US over this period, and also today, has highlighted the attractiveness of nuclear weapons for others.

Nuclear weapons are not the only, or perhaps even the main, reason for US global power, but they have played a key role in shaping the American hegemonic phenomenon. Moreover, it is also hard to refute that with one or two exepctions the other most influential states in the global system are also nuclear-armed.[7] This means that there is an intrinsic unfairness to the global nuclear order experienced by those that are not part of this exclusive nuclear club. While the nuclear-armed states may clash over the nature of the global *order* more broadly—and, for Russia, China, and others, seek to revise or push back against US/Western dominance—this doesn't necessarily equate to a shared desire to change the global *nuclear order*. All current nuclear-armed states see a role for nuclear weapons, none are realistically entertaining ideas of disarmament any time soon, and all see a benefit in limiting horizontal proliferation to others.

For sure, the Third Nuclear Age may well be characterised by an amplification of previously marginalised viewpoints and voices challenging the entrenched status quo, and potentially a greater role for non-nuclear-armed middle powers, but their ability to overturn this enshrined nuclear orthodoxy may be limited. Ultimately, this is because the question of who can bring intellectual and material influence to bear on nuclear politics will be central to what is possible in the Third Nuclear Age and beyond. At least for the time being the centre of gravity for this will be with the US and its allies.

'Plus ça change, plus c'est la même chose' (or don't get too drawn to things that are new and shiny)

Not everything in nuclear politics in the Third Nuclear Age is new, and we should be careful of focussing on what appears to have changed at the expense of the dynamics that remain, albeit in some cases slightly altered from the past. The current global order does increasingly reflect changes in technology, geopolitics, and norms, but these are perhaps better thought of as gradual transitions and modifications rather than moments of rupture, and reflect shifts in threat hierarchies rather than their replacement. In this

[7] Germany and Japan are perhaps exceptions—but their populations both live under the US nuclear umbrella.

way, many of the challenges and attributes of the First and Second Nuclear Ages remain fundamental in the Third.

Perhaps the best example is that, other than getting smaller and more efficient, the basics of nuclear warheads or bombs haven't really changed since fission and fusion weapons were first developed in the 1940s and 1950s. Likewise, while the systems used to deliver these warheads have evolved and, in some cases, become more stealthy or accurate, the basics of ballistic missiles, submarines, and bomber aircraft all remain largely unchanged from the early First Nuclear Age, and continue to be seen as central to nuclear deterrence strategies today. This is not to say that artificial intelligence (AI); extremely fast, manoeuvrable, and accurate delivery systems; the cyber challenges that emerge as a result of digital interconnectedness; the proliferation of autonomous platforms; or the non-kinetic capabilities deployed in different domains are not important—they are. But this is as an addendum to technologies that already exist, that are spreading to more actors, and that are becoming more capable.

The same is true when we think about the frameworks for managing nuclear technologies. The Treaty on the Prohibition of Nuclear Weapons (TPNW) may be symbolic of a new era, but almost all signatories continue to see the NPT as central to nuclear governance and not something that should be replaced. Moreover, while the wider perception of the good-faith commitment to disarmament by the nuclear-armed states is eroding, the majority of policy and diplomatic elites in other parts of the world do not welcome the idea of further nuclear proliferation either. This fear in itself was intrinsic to previous nuclear ages, albeit with a shifting cast of would-be pantomime villains. That said, very little has changed that makes building the bomb easier today than it was in the 1960s or the early 2000s, in fact, it may even be more difficult than in the past due to increased surveillance and monitoring.[8] The development of 'civilian' nuclear expertise as a pathway to weaponisation remains a concern, but so does the desire for others to reap the rewards of nuclear technologies for non-military purposes that has been a consistent theme across the nuclear ages. While north-south debates about nuclear energy were a central feature of the 1970s, this has arguably become much more conspicuous today.

[8] This is discussed in David M. Allison & Stephen Herzog, 'Artificial intelligence and nuclear weapons proliferation: The technological arms race for (in) visibility', *Risk Analysis* 45:11 (2025), 3839–59, https://doi.org/10.1111/risa.70105.

The split between the nuclear-armed actors and those demanding a change to the nuclear order has been central to the nuclear story since its inception. But while the nuclear order has always been contested and reflects a clash of ontologies—that is different visions of how the nuclear world does and might exist—with a few exceptions, a particular version has dominated: one where the most powerful have nuclear weapons, their spread is restricted, disarmament seems unlikely, and everyone else is kind of expected to accept it. This reality may not be swept away in the Third Nuclear Age, but divisions are becoming more pronounced, and different voices are increasingly having an impact on the global debate and public consciousness. Moreover, looking a bit further over the horizon, a shift in the political geography, generational change, and demographics of nuclear knowledge production (and absorption) away from the Western world could be a harbinger of change ...

This is perhaps why 'new' is not quite to correct adjective for today's nuclear world: some things clearly have changed, but a lot of things haven't or at least have evolved—albeit at different tempos for different actors in different parts of our global society.

Always be prepared ...

In 1956, a committee led by Sir William Strath published a report into the challenge of protecting the UK against a limited nuclear attack by thermonuclear weapons. The report, which was subsequently classified by the then British government, painted a terrifying picture of what just 10 Soviet hydrogen bombs would do to the UK, and set out an equally terrifying—at least from an economic point of view—set of recommendations for what would be required to address this threat. According to the British historian Jeff Hughes in 2003,

> Strath argued for the transformation of the physical and social architecture of the country. He also proposed a massive, long term programme of public information to improve awareness of the hazards of radioactive fallout and improve the chances of individual and national survival in the event of attack.[9]

[9] Jeff Hughes, 'The Strath Report: Britain confronts the H-Bomb, 1954–1955', *History and Technology* 19:3 (2003), pp. 257–75.

The UK would make some attempts at 'civil defence', but this was effectively shelved in the 1960s because it was seen as too difficult and too expensive. A similar conclusion was drawn in the US. While others took a slightly different approach: notably since 1962, Swiss federal law has mandated that there must be an underground bunker for every citizen,[10] most of the world has chosen to ignore—or at least quietly abandon—the issue of civilian protection in the event of a nuclear war. The same is not true of efforts to ensure the 'Continuity of Government' after a nuclear attack, which has received considerable attention and funding ...[11]

Decades later, we are again returning to a world of thinking about preparing societies for nuclear use. This has been driven by the return of the likelihood of conventional war, including between and involving nuclear-armed states in Europe, south Asia, northeast Asia, and the Middle East, and simply because you don't need many nuclear explosions to destroy a lot of cities, kill millions of people, and obliterate the way that societies (and the modern globalised world) function. A similar concern arises regarding the ability to respond to a radiological disaster, most likely as a result of a meltdown at a nuclear powerplant, and especially if, as seems likely, such technologies spread to more operators across the world. While by no means a perfect analogy, the response of governments and citizens to the 2020 'COVID-19' pandemic may provide a largely unnerving glimpse into how well prepared most governments and societies are to dealing with major disasters.[12]

To be fair, some governments have begun to respond to this pressure: Finland has an extensive system of bunkers;[13] in 2024 the Swedish government issued every citizen with a copy of the pamphlet, 'In case of an emergency

[10] Kessava Packiry, 'Switzerland gets "gold standard" for designing bunkers', *SWI*, 29 June 2023, https://www.swissinfo.ch/eng/business/switzerland-sets-gold-standard-for-designing-bunkers/48604892.

[11] See, for example, Garrett M. Graff, *Raven Rock: The story of the U.S. Government's secret plan to save itself, while the rest of us die* (London: Simon & Schuster, 2017); and Duncan Campbell, *War plan UK* (Ashford: Paladin, 1983). See also, the hit TV show *Designated survivor*.

[12] Andrew Futter, Samuel I. Watson, Peter J. Chilton, & Richard J. Lilford, 'Nuclear war, public health, and the COVID-19 epidemic: Lessons for prevention, preparation, mitigation, and education', *Bulletin of the Atomic Scientists* 76:5 (2020), pp. 271–6. See also Jeffrey Lewis & John Krzyzaniak, 'How the coronavirus outbreak is like a nuclear attack: An interview with Jeffrey Lewis', *Bulletin of the Atomic Scientists*, 20 March 2020, https://thebulletin.org/2020/03/how-the-coronavirus-outbreak-is-like-a-nuclear-attack-an-interview-with-jeffrey-lewis/.

[13] 'Finland's colossal bomb shelters a model for jittery Europe', *France24*, 4 April 2025, https://www.france24.com/en/live-news/20250404-finland-s-colossal-bomb-shelters-a-model-for-jittery-europe.

or attack';[14] the Netherlands issued something similar in 2025.[15] While these are eerily reminiscent of the 'Duck and Cover', 'Protect and Survive', and 'Burt the Turtle' initiatives of the distant past,[16] it is perhaps a more honest engagement with their electorate about the realities of the world in which we live.

While physical protection is difficult and very expensive, intellectual preparation is theoretically cheaper, but trickier. Raising 'nuclear IQ' is clearly important in order for more people to engage with and understand nuclear issues; but my fear is that general comprehension is low and often incorrect—and what represents raising 'nuclear IQ' is often in the eye of the beholder and guided by what a government/agency might wish people to think. By way of an anecdote, in 2022 I was asked by a prominent journalist about how the 'UK could shoot down Mr Putin's missiles'—to which I responded that we (living in Leicester at the time) can't; the UK does not have an active national missile defence system. My interlocutor, working for a respected global publication, was shocked—there seemed to be some belief that this is what 'nuclear deterrence' meant.[17]

For sure, raising the nuclear IQ can be difficult due to two things. First, the somewhat cynical view is that there may be vested interests that don't want citizens to know just how vulnerable they are to the effects of a nuclear attack. As former US National Security Council advisor Dallas Boyd has argued, 'it is hard to convince its cardinals to educate the public about nuclear strategy when their status and livelihoods depend on its incomprehension.'[18] Second, more prosaically perhaps, is a balance between education and hysteria, and how to shape the message that citizens across the world are effectively hostages to this, and that no defence exists. Both, I would suggest, are particularly acute questions for democracies.

[14] 'In case of an emergency or attack', Swedish Civil Contingencies Agency, November 2024, https://rib.msb.se/filer/pdf/30874.pdf.

[15] 'Dutch households to receive manual for surviving first 72 hours after national emergency', 18 November 2025, NLTimes, https://nltimes.nl/2025/11/18/dutch-households-receive-manual-surviving-first-72-hours-national-emergency.

[16] As Burt memorably put it: 'There was a turtle by the name of Bert. And Bert the turtle was very alert. When danger threatened him he never got hurt. He knew just what to do: He'd duck and cover! Duck and cover! He did what we all must learn to do. You, and you, and you, and you: Duck and cover!'.

[17] Versions of this interaction were regularly repeated with other reporters over the next few years of the Ukraine war. It is also a story that I have heard informally from other friends and colleagues that have been approached for comment.

[18] Dallas Boyd, 'Challenging nuclear bromides', *Survival* 65:5 (2023), pp. 75–94.

As a 'generation Y millennial'[19] growing up in a smallish village in the UK, I can honestly say that I wasn't aware of nuclear weapons for most of my early life—unlike my mum who even now vividly remembers the nuclear scares of the early 1980s. It was only really after being lucky enough to get a place to study Politics and International Relations at the University of Birmingham in 2003 that I got a first glimpse into what nuclear weapons really meant. But, even in my early 20s I still had just an inkling of what nuclear weapons could do,[20] how many existed, and what a nuclear war could mean for me and my friends and family. I suspect this is true of most people, in most countries, in most parts of the world ... today.

Are we feeling lucky ...?

Dystopian futures, often involving near-total human extinction and fundamental climatic effects as a result of nuclear technologies have been a staple of science fiction for decades. Charlton Heston's realisation that humanity had destroyed itself in *The Planet of the Apes*, which hit cinemas in 1968 (sorry for the spoiler)—the same year that the NPT opened for signature—stands as a poignant reminder nearly 60 years later.[21] But the further away we get from 1945, there is a concern that we as a species become increasingly desensitised to this possibility. Indeed, as the results of a survey by Sterre van Burren, Benoit Pelopidas, and Alexander Sorg found: 'the passage of time is reducing the memory of Hiroshima and Nagasaki ... particularly among younger people.'[22] Nuclear war and societal breakdown become things that we watch on TV or are the backdrop to a computer game, rather than being real and possible, and things that could destroy societies as we happily go about our daily lives. There is also a worrying risk that a narrative of control

[19] A label given to people born between 1980 and the mid-1990s.

[20] Even while I was studying for my PhD, I was conscious that 'doing nukes wasn't the cool thing to do' in the late 2000s—and remember being asked by at least one professor why I wanted to study nukes when the world had clearly moved on ...

[21] I am grateful to Mark Phythian for pointing this out.
Interestingly the 'alpha-omega bomb' became a subject of worship in the later film *Beneath the Planet of Apes* (1970).

[22] Sterre van Burren, Benoit Pelopidas, & Alexander Sorg, 'Nuclear memories for the future: Gaps in forgetting in European publics understandings of the atomic bombings of Hiroshima and Nagasaki', *Thesis Eleven*, 8 July 2025, p. 8. Another survey found that European publics are becoming more 'hawkish' when it comes to nuclear issues. Michal Onderco, Michal Smetana, & Tom W. Etienne, 'Hawks in the making? European public views on nuclear weapons post-Ukraine', *Global Policy* 14 (2023), pp. 305–17.

over nuclear weapons and crises between nuclear-armed states in the past has taken hold, and that this past non-nuclear use (perhaps better thought of as 'near hits' rather than 'near misses') is some sort of guide for the future.[23] In reality, we have probably been lucky that nuclear weapons haven't been used in the previous 80 years, but there is no guarantee that our luck will hold as we move into the Third Nuclear Age.

Neither does a nuclear-induced existential disaster necessarily have to come about as the result of a deliberate act—such as a war that escalates involving nuclear-armed states and their allies—but could conceivably be the result of miscalculation or some sort of 'normal accident'.[24] As the nuclear world becomes more technologically complex, integrated, and real-time, the risks of inadvertent escalation that American political scientist Barry Posen warned about in the 1980s become all too real.[25] Indeed, it was perhaps pertinent that the *New York Times Magazine* published an essay in December 2024 revisiting the *Proud Prophet* crisis simulation game run by the Pentagon in 1983, where the result was escalation to a major nuclear war.[26]

As of the (northern hemisphere) spring of 2026, there are probably about 12,000 nuclear weapons in existence, almost all of which are more powerful than those dropped on Hiroshima and Nagasaki 80 years ago. And while global nuclear stockpiles have reduced massively since the 1980s, and proliferation to more actors has—so far—been limited, this is still more than enough to kill most people alive today—either directly in war or because of the effects of nuclear use on the functioning of our planetary system. Despite various other societal challenges posed by climate change, a future deadly communicable disease, or the ever-growing task of ensuring access to food and water for a rapidly expanding global population, this means that nuclear weapons remain the only way for us to destroy ourselves as a species

[23] Benoit Pelopidas, 'Power, luck, and scholarly responsibility at the end of the world(s)', *International Theory* 12:3 (2020) pp. 459–70.

[24] Charles Perrow, *Normal accidents: Living with high-risk technologies* (Princeton, NJ: Princeton University Press, 1984). As Richard Ned Lebow put it nearly four decades ago, organizational complexity today makes mishaps more rather than less likely than in Cuba 1962. In Richard Ned Lebow, *Nuclear crisis management: A dangerous illusion* (London: Cornell University Press, 1987), p. 17.

[25] Barry Posen, *Inadvertent escalation* (Ithaca, NY: Cornell University Press, 1992).

[26] William Langewiesche, 'The secret Pentagon war game that offers a stark warning for our times', *The New York Times Magazine*, 2 December 2024, https://www.nytimes.com/2024/12/02/magazine/nuclear-strategy-proud-prophet.html.

in hours rather than decades. The control of the future of humanity therefore ultimately rests in the hands of less than a dozen individuals.[27] This can be quite hard to get your head around.

Sadly, other than nuclear disarmament—which seems far away today—there aren't any quick fixes to this problem. And despite Joshua's warning in *WarGames* quoted at the start of this chapter, we are already playing the nuclear game, and we can't wish away nuclear technologies. Arms control and risk reduction will probably make a return in some guise, the worst fears of proliferation may be limited in the short term, and mechanisms might be found to reduce tensions between rivals. We may even find a more acceptable way for nuclear technologies to spread and address various societal and economic demands in a safe and peaceful manner. But these are unlikely to be permanent fixes. Likewise, while we may well begin our journey as an extra-planetary species this century, this will not solve the nuclear-related problems that exist here on earth. We cannot escape the realities of our nuclear-armed planet. Indeed, as the American political scientist Daniel Deudney has argued, expanding the human experiment into outer space might well make them worse.[28]

But it is important to finish with some optimism. Global history shows that monumental changes for the better, which few people expected, can occur, and that luck can happen both ways: avoiding disaster but also transforming our lives. There is nothing pre-ordained about our nuclear future, or the nature and scope of the Third Nuclear Age (or whatever people chose to call it) or what follows it; it is—as the British aphorism has it—'up for grabs'. Our future doesn't have to look like the dystopian world so vividly painted by Walter Miller in *A Canticle for Lebowitz* in the late 1950s,[29] or more recently in the prolific author Robert Harris' *The Second Sleep*[30]—whereby a future society tries to understand the past through its miserable post-nuclear remains. To a certain extent, we all have some agency in defining our nuclear future. Wouldn't it be amazing if the civilization that invented the way to kill itself also found a way to craft its salvation?

[27] For a discussion of this, see Jeffrey Lewis & Bruno Tertais, 'Finger on the button', Center for Nonproliferation Studies Occasional Paper 45, February 2019, https://nonproliferation.org/wp-content/uploads/2019/02/Finger-on-the-Nuclear-Button.pdf.

[28] Daniel Deudney, *Dark skies: Space expansionism, planetary geopolitics, and the ends of humanity* (Oxford: Oxford University Press, 2020).

[29] Walter M. Miller Jr, *A canticle for Leibowitz* (Philadelphia, PN: J.B. Lippincott & Co, 1959).

[30] Robert Harris, *The second sleep* (London: Penguin, 2019).

Index

For the benefit of digital users, indexed terms that span two pages (e.g., 52–53) may, on occasion, appear on only one of those pages.